Stay blessed and
encouraged!

Rev. Heatley

GOD IS THE GOAL

The Chase for Intimacy with God

TAFT QUINCEY HEATLEY

iUniverse LLC
Bloomington

GOD IS THE GOAL
THE CHASE FOR INTIMACY WITH GOD

Copyright © 2014 Taft Quincey Heatley.

All rights reserved. No part of this book may be used or reproduced by any means, graphic, electronic, or mechanical, including photocopying, recording, taping or by any information storage retrieval system without the written permission of the publisher except in the case of brief quotations embodied in critical articles and reviews.

iUniverse books may be ordered through booksellers or by contacting:

iUniverse LLC
1663 Liberty Drive
Bloomington, IN 47403
www.iuniverse.com
1-800-Authors (1-800-288-4677)

Because of the dynamic nature of the Internet, any web addresses or links contained in this book may have changed since publication and may no longer be valid. The views expressed in this work are solely those of the author and do not necessarily reflect the views of the publisher, and the publisher hereby disclaims any responsibility for them.

Unless otherwise noted, scripture taken from the Holy Bible, NEW INTERNATIONAL VERSION®. Copyright © 1973, 1978, 1984 by Biblica, Inc. All rights reserved worldwide. Used by permission. NEW INTERNATIONAL VERSION® and NIV® are registered trademarks of Biblica, Inc. Use of either trademark for the offering of goods or services requires the prior written consent of Biblica US, Inc.

Any people depicted in stock imagery provided by Thinkstock are models, and such images are being used for illustrative purposes only.
Certain stock imagery © Thinkstock.

ISBN: 978-1-4917-3236-6 (sc)
ISBN: 978-1-4917-3237-3 (hc)
ISBN: 978-1-4917-3238-0 (e)

Library of Congress Control Number: 2014907084

Printed in the United States of America.

iUniverse rev. date: 05/28/2014

Dedicated to my promise and gift from God—my wife, Krystal

My faith in life … leads me directly to the source of life which is at once the goal of life—God.
—Howard Thurman

Contents

Preface
xi

Introduction
xiii

Chapter 1
The Goal That Led to the Wrong Chase
1

Chapter 2
Preparing for the Chase: Death to Self
11

Chapter 3
The True Chase
23

Chapter 4
The Desire
33

Chapter 5
The Other Chase
45

Chapter 6
The Goal: Intimacy with God
65

Chapter 7
Maintaining the Goal: Holiness
87

Conclusion
97

Epilogue
My Testimony of the Power of God
99

Preface

This book is a reflection of my intimate relationship with God. My intent with *God Is the Goal* is not to write an autobiography or personal memoir solely. Nor is this a self-help exposition. I highlight certain experiences in my life to provide personal, real-life examples of my understanding of how God moves in intimacy with those he loves. I reveal portions of my life so that you are able to witness my progression in my pursuit of God. It is a journey of my understanding of how to reach and maintain the goal of life. Furthermore, I use scripture to illuminate truths and explicate their meaning in relation to my experience. The events in my life and the corresponding verses of scripture are meant to be illustrative for teaching. I understand that my life is a letter of recommendation from the Holy Spirit to believe in Christ. This is why I divulge my shortcomings and the lessons to be learned from them. I do this out of a motivation to help someone believe that a life with God is the best life to live.

    I write this book because of the countless individuals who tell me that I should journal my life and share what I have learned. It is hard for some people to believe that I would leave a lucrative career in investment banking to become a minister. But I feel that being a minister of the Gospel of Jesus Christ is my purpose. Furthermore, I believe that you find intimacy with God when you pursue purpose.

    I began writing this book about four years ago. I was unsure as to how to start it, until I had a conversation with my cousin Valorie Burton, who is an accomplished author. She told me to just write. So

I would write for some time and then stop. I continued this pattern for at least three years. Then, in 2013, I had an intimate moment with God. An angel spoke to me in a vision and told me to finish the manuscript and begin the publishing process.

I begin each chapter with a Bible verse that serves as a point of departure. The scripture is directly correlated to events of my life. I then attempt to give in-depth explanation of the scripture as a teaching moment for the reader to apply to his or her life. Additionally, the ends of some chapters have what I call a *truth test*. The truth test is a set of reflection questions based on the main principles or precepts from the chapter. My hope is that the reader will take an honest assessment of his or her life and move into intimacy with God. I find that God begins to move in my life when I am honest with myself. I believe honesty serves as a sign of humility. When I am humble before God, I am open and available to experience intimacy.

I am indebted to many individuals. I thank God for my wife, Krystal, to whom I dedicate this book. She has been a source of encouragement and gives excellent advice. I thank God for my parents, Dr. Alvin Taft and Dorothy Heatley, for telling me about God and raising me in the church. By witnessing their lives, I saw firsthand a commitment to God, discipline, sacrifice, and the art of giving. I hope that I can be to my son what my parents are to me. To my sister, Adrienne, my homie, thanks for being my comedic partner. To my church families, Mt. Calvary A.M.E. Church, Hartsville, South Carolina; Emmanuel Baptist Church, Brooklyn, New York; and Ray of Hope Christian Church, Decatur, Georgia—I thank you for allowing me to serve God and you. To the iUniverse staff, thank you for helping me in the developmental process of authorship. To everyone who reads the following pages, I pray that you hear God speak in some capacity and move deeper into intimacy with God. Last but not least, God the Father, God the Son, and God the Holy Spirit, you are my source and ultimate companion. Thank you for calling me into intimacy with you. I desire to live for you. I pray that within these pages you receive the glory that is due to your name.

Introduction

Intimacy with God is to know God in a love-fulfilled, trustworthy relationship. It is walking with God in assurance that God loves, knows, and desires you. It is a close and confidential communion with a personal God. In intimacy with God, God affirms your worth and value as his child. In turn, you offer God your undivided attention, reverencing him as the object of your desire. Intimacy with God through worship, prayer, fasting, and service breeds a true, lived experience where your desire is to live a holy life—that is, to acknowledge and honor God with your life. What is more amazing to me is that God desires that we walk and pursue him every day. Think about how amazing this is. The God of all creation desires you and chooses to make himself available to you. God says, "Before you call, I will answer."[1] God desires us and says, "Call to me and I will answer you and tell you great and unsearchable things you do not know."[2]

I believe intimacy with God is the catalyst for holiness. Holiness is living a life that is set apart for obedience to God. It is a process that we, as believers, choose to participate in. A healthy, devotional life comprised of worship, prayer, study of scripture, fasting, and serving in a community with other believers who pursue the same goal to know God intimately is an aid to intimacy. You spend time with God, and in your moments with God, you develop a passion

---

1 Isaiah 65:24.
2 Jeremiah 33:3.

to live a life pleasing to him. This is accomplished through loving God through obedience to his word, will, and way. A consistent practice of intimacy through the spiritual disciplines helps foster a sense of accountability where you develop a reverential fear of God. Essentially, you are serious about the life you live because you do not desire to displease God. Therefore, you rely on him to help you eradicate the excessive sin in your life. As sinners, we come to God humbly recognizing that through the blood of Jesus and our relationship with God we are being made whole. We openly confess our sin, knowing that God is faithful and just to forgive us of our sins, cleansing us of unrighteousness (1 John 1:9). We understand that in intimacy we are accountable to God and must answer for godless behavior and speech. The result should be a change in our approach to life because we desire to please God.

I am constantly amazed at the greatness of God and how close he desires to be with those he loves. Life presents us with problems and issues for which we do not have a solution. But through intimacy we discover trust in a God who always provides. At times, God pushes us into situations where we have nothing or no one to depend on but him (see the epilogue). It is an uncomfortable place, where we meet the God of all comfort, who reminds us that we are his own and makes providence and provision for the task at hand. Experiencing God in this manner fuels passion and joy for his presence. Also, the joy of remaining in a relationship with God, knowing that you are living and moving in the will of God, produces peace.

God welcomes everyone on the journey of intimacy. There is room for you, and God knows the way in which we should travel. Did you know that God is waiting and wants you more than you want God? It is true. It is the reason why God sent his precious gift, his son, Jesus, to die for us so that we can live an abundant life in him. God desires for you to be in an intimate relationship with him, and the invitation is open-ended. I pray that you will accept Jesus as your Lord and Savior, vowing to live for him. I promise you Jesus will make the journey worthwhile. He did it for me, and since God is no respecter of persons (Acts 10:34–35), I believe that God desires intimacy with you. I pray that my story and journey bless you. I hope that God reveals to you purpose in your pursuit of him. Please say yes and walk with God. An intimate journey with God is the goal of life.

CHAPTER 1

The Goal That Led to the Wrong Chase

> There is a way that seems right to a man, but in the end it leads to death. —Proverbs 16:25

My goal was to be a millionaire by thirty. I never planned on working forever, and I sure did not see myself working past the age of forty. I wanted to enjoy life, and I did not see how having a *typical* or *ordinary* career would bring me happiness. I believed that there was one critical element, an ingredient that could enhance my then naive recipe for success—money, and plenty of it. I thought that financial wealth brought happiness. I reasoned the doctors, lawyers, architects, engineers, and bankers in my town were wealthy. People who owned real estate, businesses, and assets and possessed hefty salaries of employment seemingly had no problems or issues, at least not when it came to money. Having the ability to go where you want, when you want, without any apprehension as to the cost of an item seemed like an excellent goal for me. I quickly equated having money to having freedom and success, so that is what I set my mind and heart to possessing. Besides, growing up I vividly remember my father expressing, if not complaining, that we never had enough money as a family. I was determined to have money so, at the least; money would not be an issue when I became an adult. (Now I realize that even though my family had some financial struggles in my youth, we were extremely blessed. Honestly, my father was cheap, or at least that is what my mother says.)

My ambition for financial wealth began at a young age. I loved having money and began to discover how to make and acquire it. I always had my hand out at birthdays or special occasions. I would take any form of currency, whether change or bills. If I had money, I did not spend it. I saved it and did not let my sister or my parents know how much I had. As early as the sixth grade, I had a business. At the age of eleven, I charged my classmates five cents a game to play my Pac-Man watch. I was the talk of the class. All of the boys wanted to take part in my business venture. I was doing pretty well until they realized that I was making a good chunk of change. Unfortunately I had to shut down business when my watch went out of service.

My entrepreneurial quest for money did not die with my watch. In junior high school, I bought a bag of my favorite candy and sold the candies individually in class. I have to admit that there were times when I left campus before school started to go to the grocery store to purchase my supply (I cut school). I would sell each candy for five cents and one pack for a quarter. It was a pretty good gig to have at twelve. I became well acquainted with having money, and, what's more, I liked it. In fact, I loved it. I even developed hobbies to make money.

I heard that collecting baseball cards could prove to be a lucrative investment, so I began to collect them. I traded cards and even sold some myself. I heard that if you had a rookie card of a great player, at some time you could make a lot of money. Of course, the card had to be in mint condition. My prized possession was a Ken Griffey Jr. rookie card. He was one of the rising stars in Major League Baseball, and I hoped that having his card would bring me wealth. I continued to collect cards, but I was not receiving money immediately and soon traded in the hobby to look for something else.

I didn't have a business in high school. I focused on academics and football, in that order. I had a desire to be popular, so I placed myself within many campus organizations. But money was still on my mind.

I would do odd jobs for my grandparents, who would always pay me for raking leaves, cleaning cupboards, or rearranging the garage and attic. After football season my senior year, I was forced to get a job at the local department store. I only made minimum wage, which in 1993 was $4.25 per hour. As a stock associate, I did some of the dirtiest work (cleaning public bathrooms, sweeping, retrieving dirty diapers, and the like), all for a pittance. People did not treat me kindly, and many acted as if I was beneath them. I thought, given the amount of work I did, I deserved more than my menial wage. I needed more. I did not see myself becoming a millionaire as a stock associate at the local department store. I was so glad that I was on my way to college, where I could prepare myself to attain my goal.

In college I met a number of like-minded guys who shared my same ambition. We all wanted to get money, legally of course, and put ourselves in a nice financial position. We felt that our college was the place to make it happen. Virtually every Fortune 500 company recruited at that institution. I believed I was in the right place and on the path that would lead me to my personal goldmine. There was only one problem: I had no idea what career I wanted to pursue. Initially I thought that a career in engineering would propel me to financial security, so I immediately interviewed for summer internships. But I had no luck after my freshman year, so I went back home and worked at a temp agency doing menial jobs again. The jobs were an honest living for a full-time employee, but I was a part-time summer worker. I definitely did not see wealth in my future. So I could not wait to go back to school.

I was fortunate enough for the next year to receive a summer internship in engineering. The pay scale was definitely better than my previous summer jobs, and as an intern, my salary was good. I thought that engineering would be a good path, but I soon decided that it wasn't for me as a career. By the time I reached my fourth year, going into my fifth year as a double major in mathematics and physics, I was confused and brain-drained. I really didn't know how to reach my goal. In truth, I didn't want to start a profession in science.

Physics was challenging. I just wanted to make money. I felt that my breakthrough, or at least an idea that could aid my pursuit, came in the summer of '97, between my first and second senior year.

In an innocent and nonchalant conversation in the student center with some of my friends, one of them asked, "What are you trying to do after college?"

I replied, "I don't know, maybe graduate school in engineering, but I'm not sure ... I just want to make some money."

He looked at me and said, "If you're trying to make money, you should think about investment banking."

I was puzzled but intrigued. "What's investment banking?" I asked. My friends gave a brief explanation, and I immediately interjected, "Don't you have to be a finance or accounting major?" They were all finance majors. They were on their way to Wall Street after graduation.

"No, you just have to be smart, and it doesn't matter what your major is," one of them said.

Another said, "You should attend some informational sessions of some investment banking firms in the upcoming fall and see if you like it."

"Maybe I will," I said, "but what's the salary range for an incoming analyst?"

He replied, "Between $70,000 and $85,000 your first year."

I said, "Whoa ... that's all I need to hear. Where do I sign up?"

I did the math and realized that this was the ticket to cashing in and becoming a millionaire. I quickly updated my résumé and placed it in the pile among others in the career placement office. I began interviewing for analyst positions at virtually every investment bank, consulting firm, and financial service company that came to recruit. I had at least thirty interviews the fall semester of my final year. Because my college had alumni at virtually every firm, I thought it would be fairly easy to make it into the second round of interviews and receive an invitation to a Super Saturday interview. However, I was sadly mistaken.

Throughout the fall of '97, I heard from multiple representatives of companies that I was a smart student but didn't have what it takes to be in investment banking. As a result, I began to have more conversations with friends who had internships who were already on there way to New York. My boys Lamar and Robbie told me, "Control the interview with well-informed questions. Confuse them if you have to. Remember to always play to your strengths." I was a physics and mathematics major with two summer internships in engineering. I took their advice. In interviews I tried my best to sound intelligent, and I purposefully attempted to confuse my interviewers. I used complex, scientific terms to describe my field of expertise and technical experience. Thankfully, it worked. After multiple interviews and a few second-round interviews in New York, on Christmas Eve of 1997, I received my Christmas present. Salomon Brothers, later Salomon Smith Barney, called and offered me a position as a financial analyst.

I could see the realization of my goal taking flight. As I continued to interview with investment banking and consulting firms in the spring of 1998, more offers followed. I took advantage of free trips to New York and Chicago to interview, but my mind was set with Salomon. I was on my way and having a good time while traveling the path. After graduation in 1998 and a summer of serious partying, I began a career in investment banking, and I loved it. I truly adored the thought and feel of making and having money.

Investment banking was tough. As an analyst, I lived at the bottom of the totem pole and worked extremely long hours. I did not mind the long, intense hours of work, at times spending twenty-four hours a day at the job. The company made it comfortable, so I did not have to go home. I had a number of corporate perks such as an evening per diem, a company cell phone, and no personal debt. What's more, I had five of my closest friends who all worked in investment banking and four of us worked at the same firm. We made sure that we had fun and enjoyed spending money. We attended elite parties in the Hamptons in Long Island during the summer. We were present at concert after-parties in Manhattan with celebrities in attendance. We knew the hot spots in New York and hosted our own parties as well.

My roommate and I felt as if we had the hottest bachelor pad in Brooklyn, and we loved to entertain. Our house parties were second to none in Brooklyn. We had a number of fight parties (that is, watching boxing matches on pay-per-view), Super Bowl parties, and NBA playoff parties. We even rented out venues once a year in July to host a bash with some two thousand of our friends, who would travel from all over the country to attend. This party was to celebrate our bonuses for the year. The party turned out to be a huge success. It continued to grow in size every year, so we continued to have it. Shockingly, we discovered that this was a way to make even more money. We would each pocket at least $2,000 cash after every event, so we decided to host parties more often. Eventually we formed a limited liability corporation (LLC) and promoted parties on a more frequent basis. We were popular, or at least in our own eyes we were. GFC Entertainment (we were young, with money, so we thought we needed a name) was my crew, and we were on the social scene in NYC.

Life in the Big Apple was good. My corporate perks were amazing and allowed for a glamorous lifestyle. The company offered a thirty-dollar limit for meals if you worked past 8:00 p.m. As an analyst, working past eight was the norm. On many occasions, I left work and went straight to the night club, lounge, or bar to encounter the nightlife. In 2000, I made $110,000, and my investment portfolio was expanding. But this was just the beginning. My income was on the rise. I had about $20,000 in investments, not including my 401(k) retirement fund. I was twenty-five years of age with an earning potential to reach my million-dollar goal. I was taking first-class flights to Europe and had no problem or shame telling people what I was doing. I shopped at all the exclusive stores on Fifth and Madison Avenues and in Soho. I frequently visited the Barneys New York in the World Financial Center (before the tragic events of 9/11 in 2001). I looked out for every sample sale in the city. I was an avid online shopper and knew all of the good deals. I deeply enjoyed the lifestyle.

I remember one instance after we received our yearly bonuses, my friends and I went to one of the premier steak houses in New York City. We walked into the restaurant, and people looked at us strangely. I am sure that some felt we were entertainers. We sat down, and our waiter presented us with the menu. "Good evening," he said.

We replied, "Hey, man, what's good?"

He said, "Fine, sir, what would you like to drink?

I replied with my usual choice. "Coke for me." I didn't drink alcohol, but my buddies did. They were not as shy.

They said, "We would like some champagne and cognac." We told the waiter that we were ready to order.

When the waiter asked about an appetizer, we said, "Do you have lobster for an appetizer?"

He looked confused and said, "Unfortunately, lobster is an entrée."

We looked at him and chuckled, saying, "Well, we want lobster as an appetizer before our steak, so bring out the lobster entrée as our appetizer."

The waiter still looked perplexed but said, "Sure!" He was probably thinking about the tip. Then we each ordered filet mignon for the main course.

We were full of ourselves and ludicrous. But this was the kind of bravado and wasteful consumption that we lived with. Our philosophy was, if we could afford it, why not get it? I employed this philosophy everywhere I went. On one occasion, I walked into a luxury shoe store and bought the most expensive pair of shoes they had, because no one would assist me. What point was I making? The mind-set and mentality by which I operated in were ridiculous. But I had money and thought this was how you were supposed to conduct yourself if you had it.

I felt as if I was on my way, but internally something was wrong. Interestingly, there was something nudging me during my prideful, misguided ascension to a supposed successful place of contentment. Although I was having fun, truthfully I did not like the person I was or who I was becoming. I felt as if something was missing. My life was defined by my profession, lifestyle, social status, and the chase for more money. I felt extremely empty inside. It was a weird feeling. Here I was, doing the very thing that I set out to do and, by those standards, living a successful life. If I stayed the course, I could make an amount of money that my parents only dreamed of making. Yet I felt shallow. In my estimation, there had to be more to life than this. Working crazy hours consistently and accumulating things could not be all that life had to offer. Yes, I was prospering financially, but spiritually I was stagnant. I knew the Lord. I was a devout Christian, or at least I thought I was. I went to church consistently. After all, I was a child of the church.

I was baptized at a young age and did virtually everything that a young man could do in church. I was a church boy, and I thought that I knew God fairly well. I attended church in Brooklyn while working in investment banking and even began to tithe for the first time in my life. I had a devotional life that included prayer and occasional Bible study. However, I still felt depleted. I could not understand why, but I would often hear a voice reminding me, "You know what you have to do!" Clearly I knew who was speaking to me.

I could not waste any more time running from the true path for my life. It was the life that has been ordained for me since the foundation of the world, and not the one I chose for myself. A more intense devotional life led me to face the reality of my life through the eyes of God. I was a child of God who attended church but chased the wrong prize, going after a goal that could never bring joy and true contentment. Thus, there was a deep void in my soul, and it felt awful. I had an overwhelming sense of shame for trying to become someone that I believed God never intended for me to be. I didn't know what my friends and family would think if I walked away from it all. I had a difficult time maintaining the façade that everything was great in my life. I felt that my reputation was at stake and people would not understand if I decided to make a drastic change. The overwhelming problem was that I could not find true contentment or receive adequate rest. My level of happiness ranged from low to nonexistent.

There was nothing left to do. I had to make a change. It was time for me to mature and find purpose. Taft Quincey Heatley and my ego had to die. I tried my best to process what this death would look like and how I would handle it. Up until this point in my life, I only lived for myself. My life was about me and my pursuit for something that was fleeting, something that I could not carry with me into eternity. Who would help me? How could I explain it? Although I constantly enjoyed the company of people, I still felt very alone. Was there anyone who understood what I was wrestling with? Death felt imminent.

I didn't know who I was. I was fighting to maintain an image of a false reality because I enjoyed the manner in which I was perceived more so than just being me. The way that seemed right to me was never right for my life. Instead, that road was an ambivalent path that led to my personal death. Moreover, God accelerated my death. I was up for promotion twice, but to no avail. I was denied. My death was the key motivation and preparation for what I consider the *real* chase in life. My preparation for the chase of intimacy was death to self.

CHAPTER 2
Preparing for the Chase: Death to Self

> Father, if you are willing, take this cup from me; yet not my will, but yours be done. —Luke 22:42

Wall Street was profitable, in more ways than one. I was exposed to an entirely new world that I did not know existed. I met some extremely intelligent and passionate people. I deeply enjoyed the lifestyle as an investment banker, or at least what it brought materially. But there was something nudging at my soul. I felt empty. My professional career was comprised of a lifestyle of emptiness that was used to support an image of a man that God did not want me to be. Honestly, it seemed as if I did not fit, and I could not prosper in the way I thought I should.

Interestingly, many of my close friends thought that I had it made. They loved to celebrate my success with me. But what festered in the deep places of my soul was the neglect of a call from the Creator. The God in me was asking for something else. I was running from the calling, the true calling, God placed on my life. God was calling me to serve him as a preacher of the Gospel.

I first heard the call of God my freshman year in college, but it was something I thought I could do without. My desire was to be a millionaire by the age of thirty, and being a preacher or pastor was certainly not the way to go about it. Time passed, and in my senior year, God gave me another prod when I lost my bid to become the student government association president (a low moment in my life). I remember that Friday afternoon when my late friend Virgil Maupin, who was head of the election committee, announced that I only received 46 percent of the vote in the runoff election. I was distraught and deeply wounded. It was at that moment I heard God whisper to me, "I have something better for you to do."

I knew that God called me to preach. My mother constantly reminded me that I first informed her of this in September 1993. How I came to this point is still amazing. I was deeply troubled by a conversation regarding faith in Christ that I had with an upperclassman who told me that my entire outlook on God was wrong. He said, "Q, you have a good heart, but your approach to God is wrong."

I asked, "How could you say that?" We argued back and forth for a while, but I decided that I had enough. I left the conversation and called my mother. When she answered the phone, I wasted no time. "Ma, I think God is calling me to preach."

She said, "How do you know?"

I told her, "It's just what I feel, or maybe I'm crazy."

My mom answered, "Just pray, and the Lord will reveal what's next."

I had no idea what I was saying at the moment. These words came out effortlessly, almost as if someone was speaking for me or through me. After I realized what I uttered to my mother, I dismissed the statement and continued to pursue my own path. I reasoned that I would eventually fulfill the call, but I had other aspirations. I had money on my mind.

After all, I didn't think that I had to be a broke preacher. So after I graduated from college, I made a *deal* with God. In my prayer and devotional time, I told God that I would work for two years and then attend divinity school or seminary. This was the plan that I told a number of my friends about. The conversation concerning my life calling with those I trusted was met with many different responses.

Some accepted it with gladness. Others were greatly confused. Yet, God, in his permissive will, honored me and my proposition.

God made it possible for me to reach Wall Street, and, obviously, I was grateful. But there was a slight problem. A two-year career turned into a three-year career and later a four-year career. Divinity school was nowhere in sight. I knew that God was patient, but it seemed as if that patience was running thin. Or was I trying to bargain with God for my life? It was during my third and fourth year in the corporate world that I begin to see the omnipotent and sovereign arm of God at work. I was working against God rather than with him.

I could not prosper further in my career as an investment banker because I was working against God's will for my life. For some reason, I could never obtain a promotion to the associate position. I was promoted as a senior analyst after my first two years but could never rise to the rank of an associate. After my first attempt at a promotion was denied, I made a lateral move in the firm as a second-year senior analyst. When promotion time came again, I was rated the highest analyst in the group, but there was an analyst who was promoted instead of me. (Apparently there was already an agreement in place before I arrived to the group.) They wanted to keep me as a senior analyst for a third year, but I refused. Instead, I resigned and finally began to humble myself and recognize my reality. I knew that I could not fight God and win. God mercifully allowed me to continue down my path, only to discover that my path was not the one he intended.

It got worse. After I left Wall Street, I could not find employment. I thought that I would save my money for divinity school, but again my thought process did not line up with God's plan for my life. Eventually, I ended up working in the after-school program at my church in Brooklyn, where I would eventually preach my first sermon a year later. I loved what I did, but my internal struggle of not living my prior lifestyle was strenuous. I made significantly less than I did on Wall Street. It was a part-time job with a salary slightly above minimum wage. Because of the high cost of living in New York, I exhausted all of my savings, investments, and retirement to survive.

It was definitely the most humbling experience in my life. While my friends prospered in their respective professions, I was broke. At times I did not know where my next meal would come from or what it would be. Some nights my diet was pretzels or candy. I had to adjust my perspective. Although it was a hefty piece of humble pie that I knew was good for me, it did not taste good. I felt confused. I believed that I was on the right path but never envisioned that I would be here. My goals and dreams were not in line with my reality. Here I was, a college graduate who started his professional life with a lucrative career, now barely surviving. And this was all a result of following God. My view of how God worked was nothing like this. I was definitely in the wilderness, and now I had to experience that God was still God, even in the wilderness.

Through this experience, God taught and forced me to depend on him for everything. I could not rely on my strength, and definitely not my finances. I could not rely on the help of friends or family. I could only depend on my God to see me through. I saw the sovereign, providential hand of God in unimaginable ways. Many people that I helped at some point by loaning or giving them money now returned the favor in my time of need. And I assure you that they were prompted to do so by God. I learned that humility was the key to the death of self. I found out that God was all I needed, because God was truly all I had.

If you humble yourself before God, you will begin to discover what it means to trust God. You will experience the mysterious way that God provides and surprises his children with abundant blessings that are incomparable to any material asset. I am not suggesting that God does not desire for his children to prosper financially. I know a number of devout believers who God has blessed with a tremendous amount of wealth, some of whom work in investment banking. After all, God has given us the ability to produce wealth (Deuteronomy 8:18). The question becomes at what cost do you pursue financial security and success to the point that it becomes your only pursuit? Does it consume every ounce of your being? I believe that money should never be the sole focus or goal in life. If it is, then money becomes an idol—something that is worshipped and consumes all of your energy, efforts, and thoughts. Whatever you most often give thought to has a proclivity to become an idol.

Money was my idol. Money was what I lived for. Furthermore, because we live in a capitalistic society, it is easy to participate in the rat race to want and acquire more. We acquire more assets, more possessions, more real estate, and ultimately more money. Virtually in every facet of American culture, people are celebrated for what material objects (houses, cars, gadgets, clothes, and so on) they are able to attain. Accumulating wealth is not a guarantee of happiness and definitely cannot ensure a firm foundation in the Kingdom of God. Money should never be the sole motivation for living, and it definitely cannot be the goal of life. This path doesn't lead to God but to greed. As we travel throughout the journey we call life, let us remember the words of Jesus, who encourages us, as he did the children at the Sermon on the Mount in the Gospel of Matthew: "Do not lay up for yourselves treasures on earth, where moth and rust destroy, and where thieves break in and steal. But store up for yourselves treasures in heaven, where moth and rust do not destroy, and where thieves do not break in and steal. For where your treasure is, there your heart will be also."[3] Perhaps Jesus expresses the point more clearly and explicitly in verse 24: "No one can serve two masters. Either he will hate the one and love the other, or he will be devoted to the one and despise the other. You cannot serve both God and Money."[4]

The true treasure in life is to know God. Everything else in life is fleeting and is of no comparison to knowing Jesus Christ, who is the shepherd and bishop of our souls (1 Peter 2:25). He is the giver of every good and perfect gift (James 1:17) and is the one that we are to desire.

---

3   Matthew 6:19-21.
4   Matthew 6:24.

Notice, wealth and prosperity are not merely confined to financial terms. The term *prosperity* speaks more to a wholeness of the individual being complete. This cannot occur outside of a relationship with God, especially since God already promised that he will supply all our needs according to his riches in glory (Philippians 4:19). For anyone who pursues money relentlessly, I must ask the question, when will you realize that God is enough? The disciplines of experiencing submission and surrender to God will place you on the journey of dying to your desires and living for God. What follows is a closer look at the disciplines of submission and surrender. We take a look at a moment in the life of Jesus when he exhibits surrender to God. Jesus Christ models submission to God and God's will.

The Disciplines of Submission and Surrender

The discipline of surrender is a key to obtaining peace. The scripture at the beginning of this chapter, a prayer that Jesus utters in the garden of Gethsemane (his prayer closet before his darkest hour), is a prayer of submission. Yes, Jesus fulfills the prophecy of Isaiah both by his coming and his ministry, yet at this point in his earthly ministry, one task remains unfinished. He has an impending date with death on a cross that he cannot avoid. His time has finally come, and he can no longer evade the fact that he has to do the will of his Father—be the Savior of the world. Essentially, when Jesus says not *my* will but *thy* will be done, he subjects himself to the will of God. When Jesus prays this prayer, he pronounces his death. What is so powerful about Jesus' intense prayer in the garden of Gethsemane is that he willingly submits to the will of the Father, knowing that Calvary is waiting for him. It is perhaps one of the truest examples of death to self in the Bible.

The concept of surrender, or *death to self*, is a matter of forsaking one's will, desires, and aspirations to take up and follow the will of God. It is not asking God to bless our plans and goals. Rather, it is more of saying to him on a continual basis: God, what is it that you would have me do? How can I please you and do your will? I understand that I have desires, but I want my desires to become your desires for me. God, I know that I am yours and that I have been created to fulfill your will for my life, but please show me your will. I find comfort in the words of David, who implores, "Delight yourself in the Lord, and he will give you the desire of your heart."[5] It is the wise art of submission.

To submit means to humble oneself before God. The writer of the book of James, the brother of Jesus, encourages us to submit to God (James 4:7), because God opposes the proud but gives grace to the humble (Proverbs 3:34). The word *submit*, which the writer of James uses in its most basic form, means to place under rank or to bring under subjection. Originally it is a hierarchical term that stresses the relation to superior.[6] Perhaps its most common use was as a Greek military term, meaning "to arrange [troop divisions] in a military fashion under the command of a leader."[7] In the military, acknowledgment of rank is essential to functioning and operations. Failure to observe and obey the command of the ranking officer can result in severe and grave consequences. Just ask any commanding officer who is responsible for soldiers. One critical decision determines the course of the brigade or fleet. Those who have rank earn it and have the requisite experience to prove it.

As soldiers in the army of the Lord, we must submit to the all-knowing God, the all-sufficient commander who never loses. God has already ordered our steps if we choose to follow. Submitting means we make an authentic acknowledgment that we do not know how to navigate the treacherous terrain of life. We do not know where we should go, nor do we have the directions to get there. Even if we experience some measure of success in our travels, eventually

---

5 Psalm 37:4.
6 G. Kittel, G. W. Bromiley and G. Friedrich, eds., *Theological Dictionary of the New Testament*, vol. 8 (Grand Rapids, MI: Eerdmands, 1976), electronic edition, 41.
7 Ibid.

we find ourselves at a crossroad, fork, or dead end. To submit is to truly humble ourselves before God, tossing away the pride, arrogance, and haughtiness that plagues us and clouds our vision. Pride and arrogance will obstruct your sight, just as it did mine.

Truthfully, we do not know what awaits us down the road of life. There is no guarantee that what we have today will be with us tomorrow. Life presents enough difficulties and challenges, and I am sure that no one desires to have more. I firmly believe that submitting to God is the first and vital step to success and elevation in his kingdom and, therefore, in life. Every believer must submit voluntarily, as in the term's ancient nonmilitary use—a voluntary attitude of giving in, cooperating, and assuming responsibility.[8] This is the perspective and intensity with which we, as believers, should submit under God—as a voluntary act of bringing oneself under subjection to an all-knowing, all-sufficient God who knows how to live in the very world that he created. Jesus gives us the perfect example of one who was equal with God but did not consider his equality with God as something to grasp (Philippians 2:6). He humbled himself and died on a cross to give everyone an opportunity for eternal life. He set the precedent and the example that every believer should follow.

Society can encourage us, especially men, to live with pride and arrogance because they supposedly breed the attention of money and the attraction of people. However, pride is the antithesis and the main hindrance to maturing in our relationship with God. How can you ever feel that you are God's gift to the world when your life is compared to his? Pride is nothing more than a tool used to mask the face of insecurity. Only when we make ourselves known, shining a light on ourselves, can we truly acknowledge our insecurity. Why? Because what we desire is an affirmation from someone or something,

---

8   J. Strong, *The Exhaustive Concordance of the Bible: Showing Every Word of the Text of the Common English Version of the Canonical Books, and Every Occurrence of Each Word in Regular Order* (Ontario: Woodside Bible Fellowship, 1996), G5293.

instead of realizing that God already affirms us (Psalm 8:5). Having pride may seem beneficial for a while, but God has an interesting way of bringing humility into the lives of those who claim to love him. Humility is the key that unlocks the door to wisdom.

When you live and breathe with an attitude of humility, God expands your capacity to become wise. King Solomon, one of the wisest men of antiquity, once wrote, "When pride comes, then comes disgrace, but with humility comes wisdom."[9] Humility was the key to God unlocking the doors within my life, allowing me entrance into places that I could not visit on my own. The testimonies are too many to name. I have avoided heartache, pain, and unnecessary suffering because God informed me not to take certain paths, and, amazingly, I now listen.

Humility is the attitude for those who choose to submit. It is the best way to position oneself to pursue the will of God. In fact, it is the best way and the route that Jesus Christ traveled, as shown by his prayer in the garden of Gethsemane. I know firsthand the value of a humble attitude coupled with a desire to please God, because in my days as an investment banker, I did not always act with humility.

I once lived a life with no intentions of exhibiting humility. I thought that my life was all about me and what I desired. I was on a focused mission to become financially wealthy, only to discover that I was ignorant and spiritually broke. For me, submission to God was painful because I fought it with self-conceit and a severe case of arrogance. Even when I began to respond to the call God placed on my life, I still had to learn how to submit. I did not begin to understand humble submission until I relinquished my haughty mind-set, buoyed by the loss of the material items that used to support my lifestyle and ego. Submission to God was extremely painful, even more so for me because of what I thought I had achieved thus far in my life. But I now understand that my self had to die so I could take up the mantle of God.

---

9   Proverbs 11:2.

Death to self is the necessary position that you must take as you participate in the true chase after the only goal—a goal I initially did not choose but one I now gratefully choose because of a merciful God who continually teaches me the art of submission. God promised to lift me up when I humble myself before him (James 4:10), and this is what I have experienced. Submission and surrender was me dying to my will and desires. In the next chapter, we depart from my life narrative to expound on the discipline of seeking God. Now, take time to reflect on your life regarding success, submission, and surrender. A list of questions follows in the truth test after this chapter. After you reflect on the questions, pray and see in what area you believe you need God to help you grow.

Truth Test

Listed below are some questions that I frequently ask myself, given my experiences with success, submission, and surrender. I invite you to take an honest assessment of your life by answering these questions truthfully. Then pray about the necessary changes, if any, you should make to get into alignment with the will of God.

- What is most important to you in your life (personal relationships, material possessions, financial success, career success, etc.)? How is money related to it? Is money your idol?
- What is your definition of success? Is God included in the definition?
- Do you feel that the life you desire for yourself is the life that God desires for you?
- Do you have trouble with surrender or submission to authority? Is it because you desire control of your life?
- In what area of your life do you need to relinquish your will (death to self) and accept the will of God?

CHAPTER 3

The True Chase

> But seek first the kingdom of God and all these things will be added unto you.
> —Matthew 6:33 (New King James Version)

As I grow older, I find joy in the simple things in life. One simple indulgence that I have always loved is the vending machine. This may seem odd, but there have probably been many occasions when you were hungry or thirsty for a quick snack to hold you over until lunch or dinner. I have attended multiple conferences and conventions and there are times when I crave a snack or something sweet. I love to venture to the snack area to see what I can find. My favorite snack is peanut M&Ms. As simple as it seems, I become elated when I can find a vending machine that has peanut M&Ms to satisfy my hunger. However, I have been disappointed many times. On multiple occasions, I have walked to the snack machines, ready to insert my coins or dollars bills, only to find a sign that says, "Out of order." This phrase is indicative of a certain period in my life. When I was chasing greed, my life was terribly out of order.

    I was recklessly chasing after money. I thought that money would thrust me into the who's who of society and grant me security in life. I thought everything would come easy once money was not an issue. God mercifully humbled me in my quest for financial wealth and riches, when I became broke and had to change my life in every way. I realized that I was on the wrong path and my passion and pursuit were out of order.

Everyone chases or wants something in life. In virtually every facet of society there is some prize or goal that people go after. Consider, for example, the world of sports. In every sport, in every season, teams prepare for months, and in some cases years, to reach the championship game, series, or final match and become the victor. In recent years, the PGA and NASCAR have augmented the stakes of the yearly season by adding a chase, an extra incentive to be crowned the ultimate champion in the sport. The FedEx and Sprint Cups are the ultimate achievements in the PGA and NASCAR seasons. In NASCAR, every driver accumulates points for where they finish in each race. Near the end of the season, ten drivers qualify for a chance to receive a cash bonus and be rated the winner. The concept is similar in golf, in which golfers acquire points and rankings based on their performance in the tournaments. The ten-million-dollar purse for the winner of the chase is an extra motivation to finish well. It is simply known as "the chase."

Everyday life can be presented as a chase or race toward something. We always go after some goal and many times operate with an insatiable motivation—meaning that we are never satisfied and are always searching for more. In fact, our lives are inundated with the lure of progressing. There is always an ascension or graduation to another level, whether academically, professionally, and even socially, that we seek to reach. In American's free-market enterprise, it seems as if everyone chases after a certain level of wealth to achieve the so-called American dream. From a young age, boys are socialized to chase after girls, which later means that men chase after women in a catch-me-if-you-can scenario, hoping to have them as a mate or significant other. Even in the world of entertainment, there is some element of a chase. In virtually every action movie, there is a car, air, or foot chase, whereby some character aggressively goes after a villain or suspect (or vice versa). The point is that in every realm of society, with real life or in artful interpretation, we witness a chase. We, as people, love to chase something. But what are we really chasing? Are we merely following in the footsteps of those who lived before us? Are we only doing what we saw someone else do? Are we putting too much worth in things that really don't matter? And where does God fit into this complicated equation called life? What about chasing after God?

At first glance, it seems odd that anyone would chase after God. How can you chase a God whose spirit is everywhere? One of the major attributes of God is omnipresence (presence everywhere). The psalmist King David in Psalm 139 explains that in his prayer there is nowhere we can go where God is not:

> Where can I go from your Spirit?
> Where can I flee from your presence?
> If I go up to the heavens, you are there;
> if I make my bed in the depths, you are there.
> If I rise on the wings of the dawn,
> if I settle on the far side of the sea,
> even there your hand will guide me,
> your right hand will hold me fast.
> If I say, "Surely the darkness will hide me
> and the light become night around me,"
> even the darkness will not be dark to you;
> the night will shine like the day,
> for darkness is as light to you.[10]

Everywhere we go, God is there, even if we do not feel the presence of God. Our feelings have no bearing on the whereabouts of God. Even when we cannot sense God, God is still there. The prophet Jeremiah reiterates the omnipresence of God as he chides the lying prophets of Israel. The word of God comes through the mouth

---

10   Psalm 139:7–12.

of Jeremiah as Yahweh (Lord God) breaks through and speaks, proclaiming that God is everywhere:

> "Am I only a God nearby,"
> declares the Lord,
> "and not a God far away?
> Can anyone hide in secret places
> so that I cannot see him?"
> declares the Lord.
> "Do not I fill heaven and earth?"
> declares the Lord.[11]

You might be asking, "If God is everywhere, then why do I have to pursue God?" The thought of this seems foolish. It may not make logical sense, but the more I walk with God, the more I realize that he does not operate by human logic but rather through divine, mysterious ways that are antithetical and totally opposite to how I order things in my mind. After all, if God operated the way I wanted him to, who would truly be God? Surely not me! Therefore, my faith pushes me to believe and go after him, hoping and believing the reward I find is God (Hebrews 11:6). This is chasing intimacy.

The Discipline of Seeking God

The chase after God is not a matter of finding him. God is not lost. Rather, it is a matter of seeking God so that he will reveal more of himself unto us (James 4:8). There is always more to know about God because his wisdom, intelligence, and ways are more expansive than the world and universe in which we live. They are infinite. Seeking God is not a matter of gleaning information about God, but one of revelation. It is to know him personally. He is a God of revelation who

---

11   Jeremiah 23:23–24.

reveals himself to those who focus their hearts and minds on him and fulfill his will (2 Chronicles 16:9).

In the Gospel of Matthew, the sixth chapter, Jesus addresses an obstacle of life that haunts all of us on the pursuit of God: worrying. Jesus encourages and commands the crowd not to worry about the necessities of life that our God has promised to provide (e.g., clothes, food, etc.)—as many times as I heard this, it didn't become real to me until I exhausted all of my savings—but rather seek first the Kingdom of God and his righteous, and these things God will add. The Kingdom of God is God's way of operation, how he chooses to rule the world through us. In all of our dealings, it is to act justly to love mercy and to walk humbly with God (Micah 6:8).

It is how God envisions that we live. It is to participate with him in the creative process of establishing new life within others. It is to live life forgiving those who hurt us (Matthew 6:13–14). It is to have concern for the poor, needy, homeless, and those who are less fortunate (Matthew 25:31–40). It is to live a liberated life, not being held captive to anything that keeps us from knowing God. It is to worship God in spirit and in truth, authentically and not out of routine (John 3:23–24). It is to love God and love our neighbor as if we were our neighbor. It is to totally depend on God for every need and desire. It is to display the power of the omnipotent God who breathed his spirit into us. (I will speak about this in depth later.)

Seeking God is not a one-time event. Even if you proclaim Christ as Lord and Savior and have a relationship with God, you must continually seek God, because there is always more to know about him. It is the same in all of our human relationships. To say that we know someone does not mean that we know all about them. The more time we spend with them and witness them through the issues of life, the more we learn who they are. This is how they earn the title of friend, acquaintance, or colleague. Even people who pledge themselves to marriage make a covenant to grow to love and know one another. Essentially anyone who desires to be in any relationship with another human being is signing up for a quest. In this quest, both parties seek to ascertain who an individual is and what motivates that individual to live. Even if two people have some relative knowledge between them, life will present itself in such a way that both individuals will discover each other regarding what and who they are. The most meaningful relationships are a constant, never-ending pursuit of knowledge about the character, fortitude, and flaws of someone else, whether it happens automatically or intentionally.

The word *seek* in the New Testament Greek is *zateo*. The writer of the Gospel of Matthew 6:33 employs the term to mean to search for something as if it is lost. "When used in a religious sense this word first denotes the 'seeking' of what is lost … as a shepherd looks for the lost sheep (Mt. 18:12) or a woman for the lost coin (Lk. 15:8). But the same term can also be used of the holy 'demand' of God who requires much from him to whom much is given."[12] One of my favorite childhood pastimes was playing hide and seek. My sister, cousins and I love to play this game. We would play inside, upstairs, or wherever our parents were not. I knew that I had the advantage if we played in my room. In my room I knew all the places to hide and no one would find me. I would stand on bedroom furniture and fit into crevices and virtually go unnoticed. I was excellent at hide and seek when I knew the domain and territory where I played.

There are times when it seems as if God is playing hide-and-seek with us. We are the ones who cannot find him, although God is everywhere. We desperately need immediate answers to our many questions. Yet, in our estimation, God seems nowhere to be found. We cry out to God and spend intense devotional time waiting for answers or at least a sign, but all we get is silence. We worship daily, pray at the altar, and sit in eager anticipation for the sermon in church, but it seems that the sermon was for someone else and not us. The pain and anxiety of life begin to surface more powerful than ever. But, the very one who promised that when we call to him he will

---

[12] G. Kittel, G. W. Bromiley, and G. Friedrich, eds., *Theological Dictionary of the New Testament*, vol. 2. (Grand Rapids, MI: Eerdmans, 1964–c. 1976), 892.

tell us great and wonderful things is seemingly nowhere to be found (Jeremiah 33:3). Deep within our souls, the doubt that we thought was dead is now resurrecting itself. We replay scenes in our minds when we were alone and crying out for help. At times, we express the feelings of the writer of Psalm 13, who asks the bold and audacious questions that we all ask, whether audibly or internally:

> How long, O Lord? Will you forget me forever?
> How long will you hide your face from me?
> How long must I wrestle with my thoughts and
> every day have sorrow in my heart?
> How long will my enemy triumph over me?
> Look on me and answer, O Lord my God.
> Give light to my eyes, or I will sleep in death.[13]

This is how crucial life is and can feel at times. We seek an omnipresent God who has yet to make his presence known. Yet this God is the only being that can quench our thirst for him and satisfy our desire to know him.

God is a master at hide-and-seek, better than I ever was or ever could be. Maybe it is because God is playing in a room he created. The earth that God created serves as his room (after all, the world belongs to God [Psalm 24:1]), and God knows all the nooks and crannies to keep us unaware of his presence. God knows where to be so that we can never find him, even though we go to the sanctuary, the place where we expect him to be. However, I have found that this intensifies the chase to be with God. (I will explore this in the next chapter.)

Seeking God is a continuous action. This is the approach with which Jesus desires that we seek after him. The chase never stops. Once we have Jesus, we still seek to experience and know more of him. It is truly a relationship, and the most important one in life at that. I like to think of it as the unceasing quest. It never stops. If our chief

---

13 Psalm 13:1–3.

aim is to know God, then we must seek after God. The unceasing quest calls for us to constantly keep our minds and senses in a mode of expectancy, whereby at any moment God can show up and reveal himself to us. It means living a lifestyle of worship and devotion. There are times when fasting and intense praying aid the process.

In fact, the first time I went on a spiritual fast to hear from God, God met me and my faith and audibly informed me of my purpose. It was the moment that changed my life, the catalyst for my detour from the wrong chase onto the true chase. The key is to live a life of constant communication with God. I have found that one of the ways God speaks to me is through his word. The Holy Spirit is the great Illuminator, who reveals to me the truths of God when my devotion is constant and consistent. If I am patient enough to wait for God to give revelation, I find that I don't have to look far for the answer. Jesus did say that the Holy Spirit would lead us into all truth (John 16:13), and I have discovered that there is no substitute for devotion.

Walking with God is cultivating a relationship with the one whose heart every believer should desire. It is a marathon, not a sprint or dash. No relationship is able to remain healthy unless you spend adequate time together. In fact, I believe that the failure within most relationships is due to a lack of communication. A relationship with no communication or quality time is dysfunctional, if it is even a relationship at all. How can one expect to know God if one does not spend time with God?

Anyone who desires to grow with God must cultivate a relationship with God, and that entails spending significant time with him through study, prayer, and meditation, for they are all forms of worship. The believer cannot have a mere weekly relationship with God, worshipping only on Sunday and not seeking to spend time with him on other days of the week. Going after God, or worship, is a daily commitment for those who desire to know him. I have discovered that once God reveals himself, I desperately want more of him. God's presence is so assuring and secure. It is unmatched to any sensation in the world. Once you taste the presence, you fiend for another opportunity to experience God, to the point where God becomes your heart's desire. However, given the fickle nature of humanity, God often orchestrates or allows painful experiences to accentuate our desire for him. At least, that is how it was for me. In the next chapter I explain how pain fueled a desire within me to seek and hear from God. I return to the narrative of my life story to illustrate the development of my desire for God. But before you move on, reflect on the set of questions in the following truth test.

Truth Test

Seeking God is an ongoing activity, not a one-time event. See it as getting to know someone you are fond of and deeply interested in. It is about always discovering something new about God. Think about being on a scavenger hunt to get the prize. With this truth test, take a moment to reflect and answer the following questions honestly. It may help to journal your answers. Remember, you will find purpose in seeking.

- What is the most important object or person you seek?
- How much time do you spend thinking about God?
- Is seeking God a joy, a burden, or merely another item to scratch off of your list of daily activities?
- Is seeking God a priority in your day? If so, how do you show it?
- Do you find it difficult to seek God? How do you seek God (spiritual disciplines)?

A powerful resource on the spiritual disciplines is *Celebration of Discipline* by Richard Foster.[14]

---

14 Richard Foster, *Celebration of Disciple*, 3rd Edition (New York: Harper Collins, 1998).

CHAPTER 4

The Desire

> As the deer pants for streams of water, so my soul pants for you, O God. My soul thirsts for God, for the living God. When can I go and meet with God?
> —Psalm 42:1–2

As noted in chapter 2, it was difficult for me to relinquish my selfish will and pursue the will of God. It was a lesson in humility. When I initially began seeking God, a drastic, if not cosmic-like, shift in my desires spawned. I began to see God afresh, in new ways, because I finally realized the importance of a sound relationship with him as opposed to simply good religion. Religion is a set of practices based on principles and doctrines that one or more persons confess. I am a firm believer in the necessity and role of religion in society, but I believe religion, especially in the Christian context, must be shaped by a relationship with God.

Submitting to God forced me to go after him so that I could learn how he expected me to live and conduct my affairs. I have been a part of religion for as long as I can remember. However, my relationship with God was a stagnant, nonproductive, impersonal set of activities. I knew God from afar as the divine being that I feared immensely. I reasoned that at any moment God could strike me down if I did something awful or detestable in his eyes. I never took the time to understand, nor was I taught, that God is to be feared in a reverential sense, because of who God is.

I spent years in church but did not know what it meant to have a relationship. I definitely did not seek God. As an adult, my seeking was a mere weekly Sunday visit, during which I was often inattentive because I had been on the streets of New York until 4:00 am. At times I thought I was doing God a favor, because at least I showed up. Obviously, there was no humility in this attitude, thinking that I blessed the creator of the universe with my presence when I had been ordained to worship him (Psalm 8:2). Even though I perceived that God called me, I can honestly confess that I did not know him personally. To think that God would call me after ignoring God for such a long time! That would change suddenly, due to some life experiences that fueled my desire for God.

Pain, Affirmation, and the Dream That Fueled Desire

A turning point in my life came around the turn of the new millennium. On Halloween 2000, I first expressed my sense that God called me to ministry. I met with my pastor at my church to express how I believed I was called to preach the Gospel. This came to be the first of many meetings between the two of us. During this meeting, I gave my pastor my business card. I told him, "Hang on to this, because I'm going to be famous."

He chuckled and said, "Okay, I will." I was not trying to be prophetic but was walking with fleshly confidence. I thought that the meeting went well. He asked me, "How did you come to the realization that God is calling you."

I said, "Well, I have this unrest, like I know I have to do this." I did not give a clear answer and could see that he noticed. Honestly, my mind was somewhere else. I cut the meeting short because my favorite hip-hop group, Outkast, was promoting their CD release and performing live in Time Square. My priorities were out of order. I purposefully placed God's will and my will at odds.

My pastor and I continued to converse over the year. As we continued our meetings, things begin to change. In a weird way my desires were shifting. I found myself studying the Bible and praying more. I did not enjoy the nightlife as much, even though I participated. I began to enjoy church, and my friends noticed a change as well. As my desires changed, this was the beginning of God humbling me. I noticed something within me. Even though I was changing and my desire for God increased, that same desire was at war with who I had been for such a long time.

It was around this time I received notice that a job promotion was not in my future. I was deeply hurt, and my ego was crushed. Even though I knew I could not work at the firm forever, the potential to make $400,000 or more was attractive. The desire to increase my salary still plagued me. It would have made sense to take this as an opportunity to focus solely on my call and make preparation to attend seminary or at least do something different. Instead, I chose to continue a career in the financial service industry. I sought new employment opportunities that had attractive salaries. I thought I was a sure winner because of my credentials, but everyone in the industry had a healthy and impressive résumé. I wanted the money, but my drive for it was sputtering. To add further injury to insult, the woman I was dating at the time was not convinced that God called me to be a minister.

She was a Christian, and we attended church together. We prayed together and studied the word. We were active in the young adult ministry at church. She was the first girl I dated who was serious about her walk with God. I trusted her opinion. She was aware that I had a call on my life and that I was in conversation with my pastor. One night we were having a phone conversation. We were talking about ministry. We had a difference of opinion about some matter in the Bible and about my life. In the heat of the conversation, I asked her, "Do you feel that God has called me?"

She said adamantly, "I'm not convinced … I don't think so." She said, "I look at my uncle who is a pastor and see his qualities. I don't see those godlike qualities in you." Her words were a blow below the belt. They stung, and I was in much pain after our conversation. But it is interesting, if not amusing, how pain and disappointment will thrust you into the position of humility and submission.

As a result of this conversation, I began to seek God more intently, because I needed to know for sure that God was calling me to preach. Her disheartening words made me doubtful. However, this doubt fueled my desire to know God. Initially the desire was merely to receive an answer from God that I was called, but it turned out to be much more. For the first time in my life, I employed the spiritual discipline of fasting, in hopes of hearing the voice of God. I did not eat between the hours of 7:00 am and 7:00 p.m., only drinking juice and water. I would pray when the hunger pains attacked. I would pray in the morning and evening while continuing my devotion. Yet when I prayed, I heard nothing. Then everything changed the morning of May 16, 2001, at 2:25 am.

My Encounter with God

As I slept peacefully on the morning of May 16, 2001, I was awoken by a presence that restricted all of my movements. My eyes were open, but I could not see. I was literally paralyzed. I felt a power that was so consuming, it is almost indescribable. I was covered under something that made everything dark. While I could not see, I heard the deepest, most terrifying, yet somehow assuring, voice I have ever heard. As I lay motionless, I felt as though thousands of volts of electricity were running through my entire body. I then heard these exact words: *Go get my message to the masses!*

I heard the voice of God. It was a loud and affirming voice. I received what I desired and more. It was an affirmation, and I had received God. I woke up and began praising him like never before. He answered me. I could not believe that God answered me. He honored me by speaking to me even though I had ignored him for such a long time. Scripture became alive. The words of the writer of James became true: "Come near to God and he will come near to you."[15] God answered my prayers. Ever since that moment, I have yearned for that presence of God.

This presence gives comfort and guidance. On many occasions I feel the restricting power and the sweet comfort of the God who loves me. I do not feel it every time I enter into prayer. Distractions are always present when I begin to focus in to try to hear from God. But the desire is still there. I am not suggesting that everyone must have an experience like mine. I hear innumerable testimonies of how God manifests himself in the lives of his children. But I do know that everyone experiences some measure of emotional pain and stress. Could it be that your pain is an opportunity for God to get your attention so that you will go after God?

It was pain that lit the match and sparked the fire for me. I was too foolish to go after God with reckless abandonment, because I was so consumed by my own distorted picture of achievement. But God was, and is, merciful, merciful enough to meet me at my time of need.

God continually uses life trials and bad circumstances to serve as a catalyst for us to develop a desire to be with him. Once you experience the presence of a gentle whisper in your ear reminding you that you are loved or the gentle touch of a comforting hand or the warm sensation that you know heaven is now in your midst, you desire that moment to last. It creates an internal sensation of desire. It is the desire for God. It is a thirst for something that only almighty God can quench. Let's explore the meaning of this desire and thirst for God.

The Desire: A Thirst for God

> As the deer pants for streams of water, so my soul pants for you, O God. My soul thirsts for God, for the living God. When can I go and meet with God?
> —Psalm 42:1–2

---

15 James 4:8.

Thirst for God is an intense desire that we, as believers, possess because we have tasted the goodness of the presence of our God. We yearn for God and cannot live until we have him. This is the emotion and ethos of the writer of the psalmist in Psalm 42. He echoes the desire of one who desperately wishes to be with God. He expresses a yearning that obviously results from prior experiences with God. He has sought God and knows what it is to "meet" with God for sweet communion and fellowship. He is believed to be in a painful and demoralizing condition due to unrelenting, oppressive enemies. His only desire is to be in the presence that is God—the only being that can satisfy and bring comfort. The psalmist is one who had once enjoyed God's presence and is pained by the separation.[16]

The imagery is of a deer in the desert, panting for a drink of water. This water will provide the sustenance necessary for the deer's life to continue. It is hunting season, and men are on the animal's trail. The deer is gasping for air, and if it does not find water soon, it will die. It cannot live without water. So the deer must run to a fountain that is never dry, to eternally quench its thirsts. It desires an ever-flowing stream. In that same fashion, we need God because God is the sustenance our souls need to survive.

I remember the hot summer days in South Carolina when I was preparing for the upcoming high school football season. Summer conditioning was even tougher than the grueling season. Once a summer, we would practice in full pads, often in ninety- or hundred-degree heat. Besides realizing how insane that was, I remember how thirsty I became as I pushed my body to its limits. I yearned for the coach to blow the whistle and yell "break" so I could rest. I remember the anticipation of the cold artesian well water on my lips, quenching my thirst and fulfilling my desire.

Have you ever craved anything so much that felt you could not live if you did not taste it? Have you ever desired a place of solace so

---

16 Konrad Schafer, *Psalms: Berit Olam* (Collegeville, MN: Liturgical Press, 2001), 108.

that you could rest from your labor? Can you recall the intensity of your desire? Do you remember the range of your emotions before the opportunity came to extinguish the fire of your want? This is what it means to have a desire for God. Pain, trouble, and unfortunate circumstances normally serve as springboards to propel us into a position to thirst for God. However, our desire should be to live with this craving, because we have encountered the soothing presence of our souls.

The soul is the center of our thoughts, will, and emotions. The word the psalmist uses to describe the soul in Hebrew, *naphesh*, is the essence by which the body lives.[17] God is so coveted that the very nature of the psalmist in Psalm 42 cannot survive unless it finds itself in the presence that is God. The psalmist wants God and God only. His frame of reference for the presence is the temple of God, but God is not confined to a physical building. However, when God's presence is made known to our earthly temples, we desire him even more. It is challenging, if not impossible, to describe the feeling of the presence of God. But it is unmatched by any human feeling or emotion, because it is the divine love that only God can provide. Charles H. Spurgeon, the nineteenth-century English preacher who is known as the *prince of preachers*, proclaims that his desire for God greatly outweighs earthly desires, because earthy desire cannot satisfy the place in us that is reserved for God. In an evening devotional, he proclaims, "As to my business, my possessions, my family, my accomplishments, these are well enough in their way, but they cannot fulfill the desires of my immortal nature."[18] Spurgeon speaks of that place within the soul that only God can satisfy.

The desire for God is something that believers must continually cultivate. It is a matter of the soul that wants God. The wrong pursuit

---

17 W. Gesenius and S. P. Tregelles, *Gesenius' Hebrew and Chaldee Lexicon to the Old Testament Scriptures* (Bellingham, WA: Logos Research Systems, Inc., 2003), 559.
18 Charles H. Spurgeon, *Morning and Evening* (New Kensington, PA: Whitaker House, 1997), 71.

in life can cause us to be mystified as opposed to realizing the true joy that lives in the presence of God. I used to believe that joy was wrapped up in the acquisition of capital, assets, and material things. But once I acquired these things, joy did not come with them. I believe that we chase after material possessions and things other than God because our minds and realities are cluttered with fantasies that movies, commercials, and entertainment create. Essentially we try to live the lives portrayed by actors and actresses. For instance, the most dominant image for me growing up was *The Cosby Show*. For a moment I wanted to become a gynecologist and marry a lawyer, because I thought that the characters from the show where the image of success. After taking biology my sophomore year of high school, my desires changed. However, when I achieved some measure of success, I still did not have peace and joy. Not until I felt the presence of God did joy become real. Honestly, I still war within myself to have this joy, but it deepens my desire for the presence of God.

Howard Thurman, the twentieth-century theologian and mystic who served as a mentor to Dr. Martin Luther King Jr., clarifies the notion of why one should desire God and the joyous benefit of experiencing God's presence: "It is primarily a discovery of the soul, when God makes known His presence, where there are no words, no outward song, only the Divine Movement. This is the joy that that world cannot give. This joy keeps watch against all the emissaries of sadness of mind and weariness of soul. This is the joy that comforts and is the companion, as we walk even through the valley of the shadow of death."[19]

In one paragraph, Thurman virtually elucidates the encounter with the presence of God. It gives the assurance that life is worth living. The vicissitudes of life present multiple opportunities to doubt, question, and wonder if what we believe and profess as a child of God is true. We no longer need to solely rely on the words of affirmation

---

19   Howard Thurman, *Deep Is the Hunger*, 7th printing (Richmond, IN: Friends United Press, 2000), 160–61.

of others when the Most High God's manifest presence in the person of the Holy Spirit fills the void within us. We can search the world for something to provide solace, but it will be a relentless pursuit of emptiness that takes precious time away from us running after our soul's desire for God's presence. When we meet God in this moment, it is sweet to our souls, because our longing is fulfilled (Proverbs 13:19). Remember, "God does not order hungry birds to eat, nor thirsty beasts to drink. Hunger itself seeks food, as thirst seeks water."[20] Hunger and thirst come from being depleted by the chase of the world where you crave the presence of God.

If you position yourself to seek God with all your heart, spending time to cultivate a loving relationship with God, you will experience his manifested presence. You will hear him speak and assure you that your life matters. You will feel the loving communion of the one who loves you more than life itself. Then you will know and not question that God is. There is nothing like a firsthand account of an event, a personal experience that eliminates denial and suspicion. Try God and see if he fits. I promise that he will always be the right size for you.

From the moment God revealed himself to me that morning in 2001, my life has never been the same. It was like I stepped into an entirely new world, with new people. And I mean that literally. As I began to chase God and his presence, I soon discovered that there were forces that desired to prevent me from doing so. I entered another dimension. As I began the true chase, another chase was present. It was the unavoidable strategic opposition. But this chase only further accentuated the fact that I am always victorious in battle. In the next chapter we will investigate the other chase and highlight the victory of every believer. I provide true examples of my battles in spiritual warfare. Furthermore, I interpret the spiritual meaning of specific events and survey scripture to emphasize the triumphant position all Christians have in Christ.

---

20 Calvin Miller, *Hunger for the Holy* (West Monroe, LA: Howard, 2003), 111.

But first, take the next truth test. This is an important survey that deals with the desire to know God. As you reflect, think about your reactions to the painful experiences in your life. Maybe grief and sorrow propelled you closer to God.

Truth Test

My experience of pain led me to pursue and desire God. Experiencing pain can be the catalyst that leads us to God. Sometimes pain, anguish, and frustration are what God uses to grab our attention. There is purpose in pain if we are honest about what hurts us. Answer the following questions honestly and see if God is using your experiences to draw you closer to him.

- What was the most painful experience in your life? As a result of that experience, were you driven to pursue God or something else? In whom or what did you find comfort?
- Do you naturally have a craving to know and experience God? Do you believe that God wants you to experience his presence?
- What do you desire most of God (to know him or for him to grant your desires)? Do you feel that you desire God at all or not enough?

CHAPTER 5

The Other Chase

> Be self-controlled and alert. Your enemy the devil prowls around like a lion looking for someone to devour. —1 Peter 5:8

There is another chase present in the lives of those who seek God. This chase is directly opposed to the pursuit of God. This chase is something or rather someone who pursues the believer. It keeps the believer running, if not sprinting, toward God to gain an understanding and revelation regarding how to deal with this enemy. This too is an unceasing quest that the believer must learn to complete by the power and the hope of glory that lives within.

In the scripture mentioned at the beginning of the last chapter, the psalmist's longing and desire to meet with God is spurred by the fact that his enemies taunt him and viscously pursue him. He understands that he can find security in the presence of God; therefore, his pursuit for the living God intensifies. As he pursues God, there is another pursuit for his life. This pursuit is unrelenting. His enemies desire to oppress him by causing a psychological dilemma in his mind, exacerbating the already extant battle that brews. If his enemies can cause elements of doubt regarding the omnipotence and sovereignty of his God through their oppressive and abusive words, it can trigger the writer of the psalm to lose hope. When hope subsides, worship becomes meaningless and perfunctory. It is merely going through the motions, with no expectation for God to do anything. The feeling is similar to walking in circles, knowing that you are not making any progress. Yet the psalmist is determined to enter into the presence that is God and continues to encourage himself by having multiple conversations with himself (Psalm 42:5, 11). Sometimes it takes having conversations with yourself to move you to a place of assurance where you realize that God can and is able to keep you safe and be your refuge and strength (Psalm 46:1).

The dilemma of the psalmist is an episode indicative of the life of every believer who actively pursues God. There is an enemy present to distract us from fulfilling our God-given purpose. This enemy is persistent and unremitting in his chase to destroy and distract the children of God. I know this personally, and the word of God is a witness to this truth. The other chase is one where Satan, the fallen angel, comes after us to distract us from, and deter, our destiny. However, we have no need to fear. "God has not given us a spirit of fear, but of power and love and of a sound mind."[21] You have the power and authority given to you by God. It is the same power that God used to raise Jesus from the dead that lives within everyone who professes and believes that Jesus is the Christ (Ephesians 1:18–20). It took me a long time to realize who I am in God (I continue to discover more about this truth as well) and the power that God gives to those who choose to believe. I am a witness to the other chase, and it has not stopped. I spent a considerable amount of time in ignorance, but I thank God for my on-time revelation. As you will see, I have experienced the force against me whose sole aim is to deter and distract me from pursuing God with integrity.

Seven Years in Obscurity: Money, Women, and the Dark World

The moment God spoke to me in an audible voice with such profundity and poignancy, my life was altered forever. It was not only

---

21 2 Timothy 1:7, (New King James Version)

the amazement of such an overwhelming experience that changed me but also the fact that I had no choice other than to answer the call. Now that I had a more intentional and concentrated focus to pursue God, I realized that my outlook and lifestyle had to change. What I normally tolerated and entertained became distractions, and these distractions were everywhere.

It became more difficult to walk away from my original intent to become a millionaire. I could not shop as frequently as I once did. I had to curtail impulsive spending. Although my income was vanishing, I was presented with more opportunities to spend money, money that I did not have. My friends were prospering financially and expected me to match their zeal for wealth. I foolishly went shopping with them when I had no money. I simply watched them spend freely, as I used to. The temptation was strong. I was approached often about deals, sample sales, vacations, and spur-of-the-moment trips. But I could not participate.

This was problematic for my ego, but I could no longer afford my former spending habits. Eventually, I began to spend more time alone; it was all I could afford to do. I thought that it was a reasonable trade-off. But the chances to splurge unnecessarily using credit cards loomed large. I was accustomed to the Wall Street lifestyle, in particular the fashion and look. Suits, shirts, and ties were my niche, but I had to pull back my pursuit. Interestingly, spending or going after money was not the only distraction.

In April 2001 I became celibate, abstaining from sexual intercourse until marriage. Some call this an extreme mentality of religious legality, but I could not reconcile being a single, sexually active minister of the Gospel. I have seen too many negative examples of men who claim to represent God abusing their sexuality by either leading an adulterous life or engaging in an abusive use of power. Besides, God was gracious and merciful to me thus far in my life, and I did not want to misuse these privileges. I knew that I did not want to become a father without having a wife. The notion of having a child out of wedlock frightened me, especially since I was called to be a pastor.

When I was sexually active and involved in romantic relationships, I knew that God was not pleased with me. I was not sure if it was a God-conscience or just an innate feeling, but I knew that I was disobeying him. If I was going to be serious about my pursuit of God and maintain any sense of character and integrity, I had to abstain until marriage. How could I stand before God's people and preach something I was not willing to live? I had no idea as to how long this time would be or if I had the power, resolve, or self-control to hold out.

As soon as I decided to go this route, temptation became ubiquitous. I had to change the manner in which I interacted with women. I could not give off any vibes or innuendos that hinted toward physical or sexual attraction. Frankly, I had to eliminate flirting. This was, and still can be, an extreme challenge for me, as it would be for most men.

My vow to save myself until marriage seemed to make me a magnet. All of a sudden, a barrage of women now approached me in ways that I had never experienced. They were blunt and expressed their interest in the two of us participating in sexual activity. One instance, in particular, occurred on a night I returned from a social with some friends from college.

I was preparing to attend a men's retreat with my church the next morning. One of the women at the social, who I knew previously, called me when I got home. "Quincey," she said, "my roommate is going to spend the night somewhere else, and I'll be home alone. I was thinking that you could come over."

I told her, "I can't do that, and you know why." She knew that I was celibate, but she was determined to see me.

She said, "Well, I can come to your place. I mean I was just thinking ..." I was not stupid and understood what she was hinting at. After all, she called me at 2:00 am. I could hear the determination in her voice.

"It was nice seeing you earlier, but I can't see you tonight."

She would not stop. "Come on, you know you want to," she said.

"I can't!" I replied. I was a minister and living alone. I could have easily obliged her, but thank God for self-control at the appropriate time.

I was approached on many occasions, both at work and church. I officially became a part of the meat market. Everywhere I went, I felt like I was on display. All of a sudden women were not shy. I can assure you this did not happen with such frequency until I became serious about Christ. These distracting forces were everywhere. I had no idea how difficult it would be, but the idea of sex is all around us in the media, music, and other entertainment. It is a challenge for any man to abstain from sexual intercourse. A lot of men value their manhood and worth by the number of women they have been sexually active with (or how much money they make). Culture teaches men that their significance is related to the frequency of their sexual activity.

However, women were not the only distraction. Probably the most annoying and agitating happened when I tried to sleep. On more than one occasion, weird things happened as I attempted to sleep. Many times my dreams seemed too real. I felt as if I was living in another world, another reality. Many times I would fall asleep only to wake up two hours later. When I finally would fall asleep, I could sense the presence of other beings. My body would suddenly become restricted as it did when God affirmed my call, but it did not feel as comforting or assuring. I could not see or open my mouth, but it was as if something or someone was on top of me trying to restrict my movement. There would be periods when this happened almost every night. I did not know what was going on. My first thought was to scream out loud the name of Jesus, but I would feel a hand over my mouth. Because this night paralysis became so frequent, I dismissed these experiences as dreams. But I soon realized that they were too real to be dreams. One evening, I remember being restricted or pinned down on my bed. I was strong enough to raise myself, but then something pushed me back down. I was in a fight. I did not know what to think. I was confused, and to add injury to insult, it became worse as I continued to pursue God, especially when I entered seminary.

Seminary was a training ground, and the knowledge I received was priceless. I was exposed to an entirely new world. However, no class in seminary could offer the revelation I sought from God regarding this type of demonic activity. Seminary did not prepare me or help me with these midnight marauders who endeavored to rob me of my peace.

From the moment I arrived at seminary, these nocturnal animals took notice of my arrival. The visits and warfare increased, occurring on an almost weekly basis. When they began to happen, I would call on the name of Jesus and the assault would immediately stop. I looked for someone to speak with about this, but many of my seminary colleagues could not fathom this and dismissed me as weird and crazy. Yet one gentleman listened with great attentiveness. I will call him Samuel. He is a good friend who understood all too well.

Samuel invited me to his home for dinner, and I mentioned what had been happening to me. When I arrived at his apartment, some of our other colleagues were there as well. Samuel said, "I was praying to have a gathering with some brothers, and the Holy Spirit told me to invite all of you for dinner to talk about ministry." We began to converse, and he shared his testimony with us. Samuel said, "I am glad to be alive, and I know the power of God." He told us about his missionary trip in Africa. "I was in a church service, and a man who disagreed with my ministry walked up and stabbed me in my chest."

We all said, "Are you serious?" He showed us the scar.

He said, "I was rushed to the hospital. I guess it was a blessing in disguise, because the nurse who attended to me is now my wife." After he gave his testimony, he told me something that blew me away. He began to address me, "Q, I saw you walking on campus last week, and I saw you aura."

I said, "You saw my what?"

"Your aura, man," he replied. I had never heard of such a thing, but he asked, "Were you fasting?"

I said, "Yeah, I was fasting last week to get closer to God because I don't think that I'm hearing clearly."

He then said fervently, "I knew it. I knew you were a prophet."

I immediately thought, *Of course you know I have a calling. Doesn't everyone in seminary have one?* (I later found out that was not the case.) I told him my testimony, and he was even more convinced of my calling. He meant that there was a special calling on my life because I was pulled out of my career to the ministry. The rest of our colleagues looked at us with perplexed expressions during our conversation.

Samuel and I began to converse more, and he shared more of his life and how he was in spiritual warfare. What he said to me at the time did not make sense then, but it is clear now. (I will expound later in the chapter.) Honestly, I thought I was just having bad dreams or maybe hallucinating. I did not believe in demons or demonic influences. I knew that hell was a real place and I had no intentions of going. This was too much for me. But what changed my mind was something Samuel said. He told me that some of the things that were happening to me were also happening to him. "Man, I feel like I'm fighting people at night," he would say.

I immediately replied, "You too? It's crazy, because I really feel the presence of people in my room, and sometimes I wrestle with them."

He told me, "Q, you are called by God, and the devil doesn't like it. Remember that you have the authority."

I said, "Cool." I knew Samuel was right. I had authority over the beings or demons, and they were afraid of me. I just didn't know who I was.

As these nightly visitations became more common, I began to see. I could sense their coming, and God would give me a vision through a dream before they came. When they showed up, I was ready. These evil beings who took on the form of humans with physical defects never traveled alone and normally were in pairs. Sometimes they would be in the form of animals (foxes, dogs, beasts, etc.). They would try to corner me or surround me, but when they tried, I would proclaim loudly, "In the name of Jesus, flee!" They would either scatter or stop and slowly retreat. Some would try to have a conversation with me before they revealed themselves, but God would allow me to sense the atmosphere. I would raise my hand toward them as I proclaimed, "In the name of Jesus, leave!" As I said this, I would feel a jolt of power race through my body. It was the same power that I felt when God initially spoke to me. After these demons took flight, I remember seeing a bright light in the corner of the room, like an angel who was standing guard.

This kind of activity began to cease at seminary, but continued when I entered into full-time ministry. The nightly visits became more intense and vivid. Instead of trying to restrain me, I literally saw armies forming to attack me on more than one occasion. In my sleep, I would feel hands around my neck trying to strangle me, but they were effortless and powerless. It was nothing but intimidation, an attempt to instill fear. It got so intense one night that when I slept, I actually heard something or someone whisper in my ear, "You will die tonight!" Of course I rebuked it and sent whatever it was on its way. Again, these were nothing more that intimidation measures (please see my testimony in the epilogue). During this time, God placed people in my life who helped me understand why all this happened. God gave me the revelation that I was in spiritual warfare. There are too many instances to name, and for a seven-year period I simply endured these attacks, because I had no clue what they were about. But God graciously began to increase my revelation and inform me that I had the power to overcome the wiles of the enemy.

These events were not a figment of my imagination, but were real episodes of my life. In fact, they continue to happen frequently. I am a living witness that there is an enemy of our soul, Satan, Lucifer, the devil, the prince of demons, who does want us to prosper. He is not an imaginary figure but a real force of evil who desires to do harm and destroy the people of God. His chief aim is to deceive the people of God, causing them to question who they are to the point where they do not recognize their God-given power and authority that every believer in Jesus Christ has. Satan is the father of lies (John 8:44), the serpent and accuser of the children of God, who has been cast down from heaven (Revelation 12:10). The word of God is too great a witness to Satan's existence and defeat. He will go to any length to destroy you, especially if you have a calling on your life. Deception is his tool of mastery, because he is a coward who cannot handle those who truly believe. Believer, you have authority to overcome and a heavenly savior and big brother in Jesus Christ, who defeated Satan! The name of Jesus is your defense, and your weapon is the sword of the spirit, the word of God (Ephesians 6:17). Yes, you have the authority to defeat Satan. Let's investigate the believer's authority.

The Believer's Authority

> He [Jesus] replied, "I saw Satan fall like lightning from heaven. I have given you the authority to trample on snakes and scorpions and to overcome all the power of the enemy; nothing will harm you."
> —Luke 10:18–19

In Luke 10, after Jesus commissions and sends out the seventy-two for ministry, they bring back a report saying the demons submit to them in the name of Jesus. Jesus tells them that Satan, the adversary, is a defeated foe who was cast out of heaven. Jesus reaffirms that they have authority over the forces of darkness (snakes and scorpions), the malicious and dangerous powers of the evil one that are subject to the Kingdom of God.[22] The New Testament Greek word for authority is *exousía*. In its most literal form, it translates as the power of doing something or having the ability to do something. In the Gospels, it is used interchangeably with power. During the times of the Gospel in the Greek world, *exousía* was used to describe "the possibility of action given authoritatively by the king, government or laws of a state and conferring authority, permission or freedom on corporations

---

22 H. R. Balz and G. Schneider, *Exegetical Dictionary of the New Testament* vol. 2, (Grand Rapids, MI: Eerdmans, 1990–c. 1993), 11 (translation of *Exegetisches Worterbuch zum Neuen Testament*).

or in many instances, especially in legal matters, on individuals."²³ Therefore, the one who holds the authority has the right to exercise what has been given to him or her. It is analogous to a king or any ruler in antiquity who issues a decree. Once issued, the appropriate response of the king's people or subjects is to enact into law the pronouncement or ruling from the king. If not, death awaits the one who does not follow the order. This may be challenging to accept for those who live in a democracy, where the people elect public officials through a democratic process and the word or law of our president, the executive branch of government, must travel through the legislative branch of Congress. This kind of government allows for checks and balances. However, ancient monarchies or governments ruled by a king operated differently. When the king or queen speaks, issuing an order, it immediately becomes law. This is the authority that we have been given from our God over the kingdom of darkness. Our King Jesus has spoken that we have victory in him.

Today, in the spiritual realm, it means that we have even more authority and power in the name of Jesus. Jesus, the King of Kings and Lord of Lords, tells the disciples and all of those who would believe in him that all authority in heaven and on earth has been given to him (Matthew 28:18). Therefore, every demon, evil being, and spirit must submit and fall at the name of Jesus. The late Frank Hammond, known for his international ministry on spiritual warfare, exclaimed that, "All believers have the name of Jesus as a power of attorney; they can act in the authority of that Name and can get the same results that Jesus enjoyed during His earthly ministry."²⁴

Believer and child of God, in the name of Jesus, you have authority over the enemy. There is no need to fear, because every

---

23 G. Kittel, G. W. Bromiley, and G. Friedrich, eds., *Theological Dictionary of the New Testament*, vol. 2 (Grand Rapids, MI: Eerdmans, 1964–c. 1976), 562.
24 Frank Hammond, *Demons and Deliverance in the Ministry of Jesus*, 2nd edition (Kirkwood, MO: Impact Christian Books, 2007), 9.

demonic force is subject to the authority of the believer.[25] This is why those nocturnal creatures who tried to intimidate me would cover my mouth. But they heard me mumble the name Jesus, and they fled. It is important to know that demons and evil spirits are terrified of Jesus. The Gospels are replete with examples of demons who subjected themselves to God the Son (e.g., Mark 1:23–26 and Mark 5:6–13).

The same is true today. Anyone who professes Christ as Lord and Savior by faith must realize who they are in Christ. We are seated in heavenly realms with Christ (Ephesians 2:6), which means that we are spiritually higher than any force of evil that tries to deceive us in believing that we do not have authority. People of God, you have authority! Remember that you are more than a conqueror in Christ Jesus (Romans 8:37). You have the *exousia* to command every spirit that is not of God to the pit of hell. To some, this may seem eerie and peculiar, but if you are serious about your walk with Christ, you will see a change once you begin to open your mouth and confess the word of God. Yes, preach to yourself so that the demonic activity around you must cease at the name of Jesus.

If you choose not to do so, you will fall victim to the enemy's schemes, the chief of which is deception and seduction.

The Art of Deception and Seduction

> You belong to your father the devil. ... He was a murderer from the beginning, not holding to the truth, for there is not truth in him. When he lies he speaks his native language, for he is a liar and the father of lies. —John 8:44

In the scripture above, Jesus' harsh words to the religious elite who accuse him of blasphemy illustrate the nature of our enemy—lies,

---

25 John Eckhardt, *Prayers that Rout Demons* (Lake Mary, FL: Charisma House, 2008), 27.

deception, and seduction. They are under the deception of the enemy. They do not believe that Jesus is the son of God and is moving in the authority given to him by God the Father. They are unwilling to accept that Jesus is Christ because Satan is at work deceiving them, causing them to be blind to the truth. The religious elite are not able to accept Jesus' testimony because Jesus does not resemble them culturally nor possess what is considered to be an earthly pedigree. They believe that the Messiah will come from "good stock" and not from the ghettos of Nazareth. How can the Savior of the world be born into such poverty? Yet, they are too convinced by a visual perception and not the truth of the word of God spoken through the prophets of their heritage. This is the art of deception illustrated—choosing not to believe something because it is totally based on physical sight and not spiritual insight. This is why it so important for every believer to know the word of God so he or she is not deceived or totally influenced by what is seen, which is antithetical to faith. (We will explore faith later.)

From the beginning, the devil lied and began his rampage on humanity, hoping that humanity will doubt the validity of the word of God. In Genesis 3, the enemy, in the form of a serpent, craftily deceived Adam and Eve. He posed the question to Eve in the Garden of Eden in such a way as to imply that God really did not give Adam and Eve a command (Genesis 3:1). He is the master of deception. The serpent told Eve that God really did not mean what he said and planted doubt in her mind. When Eve saw that the fruit was pleasing to the eye (Genesis 3:6), she ate and likewise did Adam. They were deceived.

The enemy uses words in a way that causes us to question who we are, as he did with Eve. If we are not aware of his schemes, we will fall victim every time. However, the word of God exposes the deception and tactic of the enemy because the word of God is light (Psalm 119:105),[26] and the enemy is darkness. Light always exposes darkness. It is humanly impossible to see clearly in the dark with no

---

26 Ibid., 45.

source of light. If our physical sight is better in the light, how much better will our spiritual sight be if we are full of the knowledge of God and his word?

The enemy tried to deceive me when I made the decision to chase after God. As my life began to change, the enemy began to plant seeds in it, telling me that I could remain the same and maintain an image of one whose income was increasing. If I continued to live the same lifestyle, even though I could not afford it, I would do nothing but accumulate debt and find myself in more trouble. As long as I spent money, I would hold on to a life that would not produce growth in God but would create bondage. Thus, I would be living a lie. Furthermore, the offers from the innumerable women would only allow me to continue a sinful life, which could affect the integrity of my testimony, even before I began to preach. How could I preach and proclaim a word that I could not live? Continuing to be sexually active outside of marriage would only worsen the image and potency of me as a preacher and pastor. I would be no different from any other man. If I did not change my interactions with women, I would perpetuate the negative stereotype of preachers who are whoremongers and misogynists who salivate at the opportunity for sexual delight. I knew that many of those women were on assignment from the enemy. It was too easy, and I am not that handsome of a man to have women offer their bodies in such ways. It was the old art of deception that Satan always utilizes. However, in scripture we witness that Jesus provides us with the perfect example to handle the schemes of Satan. In Matthew 4:1–17, Jesus deals with seduction and deception. His example of triumph gives us hope that we too can overcome the devil's schemes.

The Jesus Example: Matthew 4:1-17

After Jesus is baptized and commissioned for ministry, the Spirit leads him to the desert to be tempted by Satan. Satan tries to seduce Jesus at every angle. Interestingly, the tempter comes after the fast is nearly over, knowing that Jesus' flesh is weak and craves nourishment. The enemy normally comes when he perceives signs of weakness, trying to exploit our pain. For this reason, when we experience pain, it is imperative that we sprint to God and call on his name. His name is a strong tower; the righteous run to it and are safe (Proverbs 18:10). If not, the open wounds become ample opportunity for the enemy to enter and attempt to seduce us. Jesus shows that he cannot be deceived or seduced. Every time the tempter speaks to deceive Jesus, Jesus replies with the word of God. First the enemy tries to entice Jesus to address his hunger by turning stones into bread, but Jesus replies with the word of God. Then the enemy tries to use the word of God against Jesus. But Jesus knows the word because Jesus is the word (John 1:14) who is God. Finally, the devil tries to deceive Jesus by somehow trying to give Jesus what already belongs to him (Colossians 1:16–17). Again, Jesus replies with the word and rebukes Satan. The devil always tailors the temptation to seem as if it is not a temptation but an affirmation.[27] The word of God is truth, and truth always exposes lies. This is why it is important for us as believers to know and confess the word, because the validity of the testimony exposes the lies. Remember, "Temptation is not defeated in debate and dialogue, but in declaring the decisive alternative."[28]

It is interesting that much of the enemy's schemes and tactics are telegraphic and not creative. Notice that Satan came to Adam and Eve after God gave the command regarding their purpose. Likewise, Jesus is tempted after his baptism and commission for ministry. Please do not miss the timing of the devil's coming. The precise moment you set out to fulfill the will of God is when the enemy is prone to act and attack. You are of no relevance to him if you are not concerned about God. This is not to suggest that the enemy was never present before I accepted the call, but that when I began to pursue holiness, my world

---

27 Douglas D. Webster, *Under the Radar: A Conversation on Spiritual Leadership* (Vancouver, BC: Regent College, 2007), 46–47.
28 Ibid., 61.

came under attack. I am not stating that you will only find yourself under attack if you are called. However, do know that when you become serious, seeking God for your purpose, the enemy will come to distract and deceive you. But you can always give a counterattack as you learn to use the sword of the Spirit, which is the word of God (Ephesians 6:17).

The Counterattack

> Resist the devil and he will flee.
> —James 4:7

The word of God, which is the sword of the Spirit, is our weapon of war that the enemy must bow down to and flee from. God has given every believer the authority to speak his word, a word of power that we must feast on to survive. This is why many believers pray aloud, so that the enemy can hear our attack. Remember, the enemy does not have all power nor is he omniscient. He realizes and knows who is dangerous to his measly kingdom. This is why Jesus' words in the wilderness are so powerful. Satan could not stand up against the one who knows who he is. In the wilderness, Jesus provides the perfect example of how one should resist the enemy in counterattack mode.

We do not have to submit ourselves to the work of the enemy. When we sense that we are in spiritual warfare, we can attack the enemy as well. We just have to resist the enemy. James 4:7 says resist the devil and he will flee. The Greek word for resist, *anthistemi*, means to oppose, set oneself against, rebel, or stand firm. Just as we are rebellious to those things in which we do not believe, we must be rebellious to the enemy. We resist the devil, holding our ground by speaking the word of God in our prayers. Remember God did not give us a spirit of fear, but one of love's power and a sound mind (2 Timothy 1:7). We resist the enemy by speaking the word of God, letting the enemy know who he is. He is a liar (John 8:44). God evicted him out of heaven and threw him to the earth (Ezekiel 28:17).

The devil fell like lightning from heaven (Luke 10:18). Become bold in your prayer life and preach to yourself. God encourages you to do so, and Jesus made it possible (Hebrews 4:16). You are a child of the living God. You are the righteousness of God. You are the head and not the tail. You are more than a conqueror in Christ Jesus. You have power to overcome everything you face by your testimony and the blood of Jesus. You have power to overcome because greater is he that is in me than he that's in the world. You have power to bind those things that come against you and loose those chains that try to constrict you.

Remember the words of the Apostle Paul in 2 Corinthians 10: "For though we live in the world, we do not wage war as the world does. The weapons we fight with are not the weapons of the world. On the contrary, they have divine power to demolish strongholds. We demolish arguments and every pretension that sets itself up against the knowledge of God, and we take captive every thought to make it obedient to Christ."[29]

The word of God is strong enough to defend itself against the devil's schemes. This is how you win the battle in your mind as you try to make sense of your circumstance and predicaments. Speak the word and see what happens when you sense evil and danger around. (My testimony in the epilogue gives light to this truth.) Remember, you are not fighting a physical enemy in a physical battle but a spiritual enemy in a spiritual battle, even if there are physical manifestations of it. This is difficult to grasp because we are physical beings, but it is truth. When you are in the body of Christ you wrestle against other forces. The Apostle Paul says again in Ephesians 6:12, "For our struggle is not against flesh and blood, but against the rulers, against the authorities, against the powers of this dark world and against the spiritual forces of evil in the heavenly realms."[30] As demented and mean as we view some people to be, there are forces and realms of

---

29 2 Corinthians 10:3–5.
30 Ephesians 6:12.

influence that may cause them to act a certain way. Seek to always discern situations and circumstances before they arise. If you do not, you are liable to fall victim to unbelief. It is through unbelief that the enemy enters through the doors of your life to frustrate and confuse you even more.

If you seek God for his perspective, the Holy Spirit will make it abundantly clear, illuminating this truth in your life. God never desires for you to be in the dark, and God is not the author of confusion but of peace. That is why Paul further elucidates the fact that we are to put on the full armor of God to fight from a victorious position, in which we do:

Therefore put on the full armor of God, so that when the day of evil comes, you may be able to stand your ground, and after you have done everything, to stand. Stand firm then, with the belt of truth buckled around your waist, with the breastplate of righteousness in place, and with your feet fitted with the readiness that comes from the gospel of peace. In addition to all this, take up the shield of faith, with which you can extinguish all the flaming arrows of the evil one. Take the helmet of salvation and the sword of the Spirit, which is the word of God.[31]

Paul gives this analogy of a soldier's armor as he sits in jail awaiting his trial in Rome. The Roman soldier assigned to him now becomes an illustration for the church. Every piece of army represents the tenets and virtues of our faith that we must believe as we fight the good fight of faith. The truth that Jesus is the Christ holds up our garment. Because of Jesus, we are the righteousness of God that covers our core (Romans 3:21–22). We walk in peace because he himself is our peace (Ephesians 2:14). We protect ourselves with faith, knowing who we are and who God is. Our heads and minds are covered with the fact that we are saved by his grace (Ephesians 2:8). Our offensive weapon is the word of God (Hebrews 4:12). In spiritual warfare, you must know who you are in Christ. We confess

---

31   Ephesians 6:13–17.

the word of God because it reminds us of who we are and tells the enemy who Jesus is. Speaking the word reinforces the fact that we are to be doers of the word as well as hearers (James 4:22). The ultimate way we show that we are doers of the word is through holiness, a potent weapon against the enemy, the manifestation of true faith. Holiness is a product and should be the resulting effect of the goal of life—intimacy with God. We'll survey this goal in the next chapter and elucidate intimacy through an excavation of scripture references. I highlight moments of close communion with God to illustrate the reality and truth of intimacy in a relationship with God. But first take the truth test regarding spiritual warfare and reflect on the spiritual challenges you may have encountered in your life.

Truth Test

The enemy tries to deceive everyone who believes in Christ. He will use anyone and anything to distract us from pursuing God and living a life of righteousness. The enemy loves to exploit our weaknesses and test our strengths. You must know yourself and know the word of God to fight the enemy. Remember, the devil does not deal in truth but is a liar. When you operate in truth, you defeat the enemy. When you lie, you speak the enemy's language.

- What are your weaknesses and struggles in your walk with Christ? Have you noticed patterns in how you are challenged in these areas? (Pay attention to when you are tempted—the enemy may be present.)
- You are most powerful when you are consistently praying, worshipping, and seeking God, especially studying the word. Have you noticed that you struggle the most when something has your attention other than God? Watch out, because in those moments, the enemy is near.
- Do you ever feel the presence of something around you when you sleep, particularly after midnight? If so, begin to confess the word of God aloud.
- Do you struggle with believing what God has spoken to you or promised you? What are the lies that you are tempted to believe? Have you noticed that when you receive revelation

or a promise from God, the enemy shows up to deceive you and rob you of your belief?

Remember that the word of God is your sword in your armor (Ephesians 6:10–18). In Matthew 4, Jesus defeated the enemy with the word. Know the word to expose the lie of the enemy. A powerful resource of prayers from the word of God to defeat the enemy is *Prayers That Rout Demons* by John Eckhart.[32] It has helped me tremendously in spiritual warfare.

---

32 John Eckhardt, *Prayers that Rout Demons* (Lake Mary, FL: Charisma House, 2008)

CHAPTER 6

The Goal: Intimacy with God

> My prayer is not for them alone. I pray also for those who will believe in me through their message,[21] that all of them may be one, Father, just as you are in me and I am in you. May they also be in us so that the world may believe that you have sent me. I have given them the glory that you gave me, that they may be one as we are one. —John 17:20–21

We all have a longing to be loved. It is as natural as the moment a child enters the arms of her mother for the first time. The new baby that has been attached, connected, and cultivated in the womb continues to desire the warmth and love from which it came. What is more, the mother desires to reciprocate this kind of love, continuing to nurture and care for the newborn child. The two remain in their state of inseparability, a condition that began in the womb. I witness this as I see my son and wife interact, especially at his birth. I could see that when she looked at him and held him in her arms for the first time, she immediately loved him. There was no question or doubt about the love she had for her son. Love was the first and only response that she had for him. I even feel that same intimacy as I hold him. There is a connection in knowing that I am responsible for someone and must nurture them to maturity.

This kind of intimacy that parents and their newborn child share is so amazing to me. I was even amazed as I watched my wife interact with my son in her womb. Words are not necessary in this encounter, yet communication occurs. The mother may talk, sing, or enjoy the moment silently, yet she speaks with her child. Even if the child is unable to comprehend or respond verbally to what is said, the mere fact that the presence of the one who is responsible for their arrival indicates love. It is no question that it is a human manifestation of a divine reality and intent. This love and intimacy has to come from a divine source. God our Creator exhibits an extension of this love through this manner. Love and intimacy have their beginning in God.

When you think of intimacy, what immediately comes to mind? Is it you holding someone you love or them holding you? Do you think of taking a long walk with the person you adore? Do you think about the quiet and stillness of a blissful state of mind? Is it the way you feel when someone expresses the love they feel for you, allowing you to release all fear, concern, worry, and doubt? Is it when you trust someone so much that your wall of defense crumbles, your soul is bared, and your emotions are vulnerable? Is it protection and security that brings intimacy?

Intimacy is the state of being closely acquainted or familiar; a very private and personal moment. It is when two individuals or beings exist in close proximity to one another, either physically, emotionally, or spiritually. Many people think of intimacy in purely physical, romantic terms. Many view intimacy through the lens of popular American culture and entertainment, which deeply influence our definition and understanding. Given all of the sexual implications and overtones prevalent in today's music, coupled with the engrained lustful nature of Hollywood, our view of intimacy is tainted. Unfortunately, some choose to reduce intimacy to a purely sexual encounter that occurs mostly outside of the divine and sacred covenant of marriage. Because we are constantly inundated with these unjust images of intimacy, it can be extremely complicated to comprehend intimacy with God. However, intimacy with God is totally different. It is the goal of life.

God's original intent was for humanity to walk closely with him at all times. It is a truly intimate yearning to depend and communicate with God every day, at every moment. We see this theological truth evident in the relationship with God and creation in the Garden of Eden. In the Garden of Eden, there was no separation from God. Adam and Eve experienced true intimacy with him, being naked and unashamed before him, with no inhibitions (Genesis 2:25). Adam and Eve were so intimately woven with God that the two could hear him walking in the cool of the day (Genesis 3:8). But their sin caused separation that only God could mend. Through the life, death, and resurrection of Jesus Christ, God has restored the possibility of intimacy to all who believe. God the Holy Spirit, who is our comforter and counselor, desires to walk closely with us as we journey through life. Our comforter aids us in becoming one with the Father.

It is amazing to think that the God of creation wants you—yes, you. God wants to care for us and for you to experience him personally, leaving no doubt as to his existence. Worshipping God is being in sync with one another. It is like being able to finish the sentence or the thought of someone who is speaking because you know them. It is recognizing the footsteps or voice of someone you love.

One of my closest friends expressed to me how in tune and in sync he is with his wife, explaining the beauty of marriage and its intimacy. He was at a gathering with friends, and he heard footsteps above him and knew it was his wife. Sure enough, when he turned to look up the stairwell, he saw his wife coming down the stairs. I can discern my wife's laugh and voice in a cacophony of noise. One Sunday after church, I was greeting the worshippers as they left. There were many voices, and it was noisy in the sanctuary. I heard this laugh all the way across the sanctuary. I knew the laugh because that laugh has been close to me. It was the laugh of my wife, someone who has spent intimate time with me. These examples are simply those of this oneness, intimacy between two people who love one another.

This is Jesus' prayer in John 17. His desire is for all who choose to believe in him to seek and experience the oneness of the relationship that he had with God the Father. It is this kind of intimacy, this lifestyle and concept of oneness with God the Father, that Jesus exhibited as God the Son walked the earth.

The prayer at the beginning of this chapter in John 17 is truly the Lord's Prayer. Jesus, on his way to the garden of Gethsemane to begin his passion and fulfill his earthly duty to bring salvation to humanity, ends a lengthy, but intimate, discourse with his disciples with a loving prayer to the Father. Jesus prays—for himself, his disciples, and then for you and me. In his prayer for those who would believe the message that he is the Son of God, Jesus asks that all believers in him be one with one another and says that he is one with the Father. This concept of oneness is the intimacy between the Father and Son. A love that still remains mysterious.

True intimacy expresses no division. This notion is a mystery to us just as the mysterious relationship between the Holy Trinity. There is no human alive who can fully comprehend how God the Father, God the Son, and God the Holy Spirit can exist as one. In our human thought we describe them as one moving part with three indivisible units. There is no separation, division, or estrangement within the Trinity. The Trinity is one. We catch a taste of this mystery when we commit and devote our lives to God in loving service. It is a level of intimacy that we reach that causes us to yearn more for God. It places our lives in a different light, a perspective that causes us to value the true gift of life—God. Constant communion with God leads to an outcome of a loving intimacy where heaven continually meets us on earth. And the glory of God that is Christ in us, which is the hope of glory (Colossians 1:27), manifests itself in our lives. The inseparability of oneness with God is the goal of life. We have a witness and example—the earthly ministry of Jesus.

God the Father and Jesus are one. Jesus is adamant and clear about this in his testimony regarding his earthly ministry. Jesus spoke the words of the Father and did his work without complaint.

It was a work that eventually led to his death, which was all a part of God's redemption plan for humanity. Yet, our Lord is clear. Jesus told his doubters that "the Son can do nothing himself; he can do only what he sees his Father doing, because whatever the Father does the Son also does. For the Father loves the Son and shows him all he does."[33] From these words, we see Jesus' motivation for his earthly ministry. Every miracle, healing, sermon, and word and, ultimately, the crucifixion was motivated by love for his Father. Jesus only did what the Father saw as good in his sight because the motivational fact was love. That divine love between God the Father and God the Son is a mystifying level of intimacy where with everything the Son does, the Father is pleased, because the motivation of the act comes from a foundation of love. Yes, we too know the love that the Father has for us (John 3:16); it is the same love that God the Son exhibited to us. Remember the words to his disciples (and us): "Greater love has no one than this, that he lay down his life for his friends."[34] This is the nature of intimacy. God is pleased to reveal to us this intimacy when we place ourselves in the position to receive it. However, we must set aside time and have our hearts set on God to experience it.

The goal of life is to have, receive, and experience life with God, which is essentially intimacy with him. Intimacy is so crucial to our being. It is an extension of who we are as humans. We are creatures who need and yearn for intimacy. We try to find it through other means but come up short. This is why we chase dreams, money, the opposite sex, and careers—to find a sense of satisfaction and affirmation. Once we achieve these things, we discover that there is still something missing. What is missing is intimacy with God. We find that there is still something within us that longs for acceptance, acceptance that only comes from experiencing the unconditional love of God through Jesus Christ. This is intimacy. Even in the field of medicine, doctors encourage parents to hold their newborn babies

---

33 John 5:19–20a.
34 John 15:13.

so the parents can feel the presence of love as they develop their sense of self. Intimacy is vital for survival, especially as Christians in a dying world. The words of Jesus provide medicinal comfort and encouragement for the soul that breeds life.

Jesus said, "My words are Spirit and they give life."[35] Devotional time with God and God's word provides the proper environment for the manifestation of intimacy. This is true worship, whether in a private or public place. Time seems to slip away from us when we spend time with God on a consistent basis. It seems as if there are not enough hours in the day or night to work, let alone spend time with God. This time with God is so crucial to intimacy. I often feel empty when I have neglected spending time with God in the morning or evening. This is the void that I have. Everything in life could be great and the way I exactly want it, but there is still something missing. I feel powerless, ineffective, and irrelevant. There is a certain level of vigor, power, and motivation that is absent when my prayer time and devotion is short. However, when I am persistent and consistent in prayer, it seems as if that power is present with me. Perhaps this is why Jesus stole away from the disciples to pray: to recharge and refuel (Mark 1:35).

Having Personal Intimacy with God

I honestly feel that I stumbled across knowing the intimacy of God. It was not something that I was taught. The trials, pain, and confusion of life, especially regarding my purpose and calling to be a minister, brought me to a place of wanting to hear from God. I wanted to be able to recognize when God was speaking to me. When I desperately desired to have an answer from God, he did not disappoint. But it was the feeling that was so loving that succeeded the answer from God. This feeling is unmatched by any worldly pleasure. Wanting to hear from God spawned a craving for him. It is very difficult to place in

---

35 John 6:63.

words the intimacy that you feel with a loving God who shows His loving care for me on a consistent basis. It is both a feeling as well as a knowing. It is the ability to recognize His presence and purpose. Words simply cannot give credence or explanation to the feeling.

By no means am I claiming to be an expert. I don't know if someone can be considered an expert in intimacy with God. No matter how much we choose to be with God, he has the autonomy and choice to come close to you. Intimacy is initiated by him and can occur at any time. There are multiple instances where I miss God because of doubt and fear. Sometimes I entertain too many opinions of others who do not share my beliefs and concerns. I get caught up in small things and situations that are nothing more than traps to deter me from my God-given purpose. Some people choose to remind me of who I used to be. Their mission is to attempt to keep me in the bondage of who they perceive I am. But the distraction does not always come from others. Truthfully, at times I am the distraction.

In every born-again believer there exists the residue of his or her old life, lying dormant inside. I think that I have dealt with past hurts and wounds, but life has a way of exposing my slow progress, or the lack thereof. I sometimes find myself fighting myself. Often God will place you in situations that show you who you really are. But what is so amazing about God is that he will use these moments to drive me to a place of desiring more intimate time with him. These times when God reveals to me the shame of who I am interestingly propel me to a place of prayer where God's presence is so strong and the inexpressible feeling of love saturates my being and domain to the point that I have no doubt that God is alive and well. God reveals that shame is not necessary because he is still with me. God has not left or forgotten about me. These intimate moments do wonders for, and in, my spirit. They provide me with the confidence and assurance of knowing that I am loved unconditionally. There is no amount of money that can purchase this kind of security in God. It is truly supernatural and amazing to have the love of God through Jesus Christ, who does not reject us because of ourselves. But this God willingly accepts us and is patient with us as we pursue him.

The Journey of Intimacy

Intimacy begins once you realize that you are loved by God, that, no matter what you do, God does not stop his love for you. It is the reason why the Incarnate God in Christ was revealed. God is love and always will love. This truth helps me in my intimate moments. There have been times when I have not been proud of my actions and conduct. I know that too many times I have not represented God like he expects me to. Yet, God still chooses to see about me and let me know that he is there. Perhaps the most intimate moments are the times when the sweet whisper of God the Holy Spirit reminds you that he loves you.

God's love is the affirmation that we thirst for. This is why people try so hard to fit in and be accepted. Everyone desires to know that they matter and have value. They crave feelings of worth and merit. I am a living example. This is why I tried to find happiness through wealth and notoriety. I was on the wrong chase. What I was really after was for someone to notice me and recognize me for what I had. This was a pursuit of emptiness. Once I embraced the love of God, my life changed, even though the struggle for acceptance remained.

Intimacy with God reaffirms that human life matters to him and that we cannot live without the loving intimacy of our Creator. Knowing and experiencing intimacy with God will curb our appetite for seeking acceptance and approval of others, illuminating a new life encounter that provides focus and direction. Once the believer understands that she is loved intimately by God, she sees how to love herself and her neighbor. In a supernatural way, I believe that intimacy with God can deliver people from various addictions. I know this to be true from the countless testimonies I hear from those who surrender their lives to the love of God. Often people venture off into addictions and substance abuse because love and attention from others is nonexistent or sporadic. But the love of God is consistent and available for anyone who believes, accepts, and pursues it. Brothers have testified to me that when they finally understood God loved them, they did not need powder, crack, alcohol, or sex to validate them. They were able to break free. They learned that they were loved, accepted, and affirmed by God. They grew as their relationship with God grew and experienced intimacy with him. This is the goal of life.

An Illustration of Intimacy with God in My Life

I have had some intimate moments with God over the years that I would like to share. As I experienced my own struggles in life, especially in preparation for ministry, trying *to fit in* or be accepted, I had some pretty intense conversations, if not arguments, with some of my colleagues. Truthfully, I wanted to be accepted. I received hurtful darts and accusations regarding my beliefs and understanding that pierced my soul.

In ministry, everyone has their views concerning truth, especially the validity and truth of the Bible as the word of God. With people who can craft intelligent arguments, these truths come out. Many of my colleagues based their truth from their experience, as most people do. The lens through which we view scripture often shapes our interpretation. Many of my classmates had deep issues with the Bible, or at least how the Bible has been used. When it came to my interpretation, I believed, and still do believe, that the Bible is the guide through which we watch God interact with humanity throughout a long history. I believe that it is the word of God. It is a guide for daily living and drawing closer to him. For me, it is not a mere collection of ancient writings that has no present validity. No, it is the word of God that, through the teaching and wisdom of the indwelling Holy Spirit of the believer, can lead anyone to the truth that is God.

Believe it or not, I was attacked for believing in the Gospel and the foolishness of its salvation message (1 Corinthians 1:18). I could not understand how people who were preparing to be ministers could discredit and trash another's experience as well as the sacred text. I would rack my brain trying to understand the ridicule and chastisement I received from brothers and sisters preparing for ministry. I believe in the power of God through the preached word and do not relegate the miracles of God to ancient times. If God is the same yesterday, today, and forever more, surely miracles are still a vivid reality and part of our present existence. But some of my colleagues begged to differ. Again, their experience shaped their truth as did mine. I was not totally innocent. God was teaching me to have a sensitivity to other voices and experiences that differed from my own. Everyone has their own path and journey to travel, and we must learn patience and the art of respectful disagreement.

Admittedly, I was partly to blame for some of the arguments and misunderstandings. I wrestled with an idol of significance because I yearned to be heard. Wanting to be heard was an issue that extended from childhood. As the youngest child in the neighborhood, I felt as if I really didn't matter. I was just a tag-along to my sister. (Intimate time with God revealed this to me.) After the arguments in class, I would return to my dormitory, pray, and ask God to grant me revelation or truth regarding my life. Nothing warmed my heart more than hearing God whisper to me that he loved me. I have heard God whisper on multiple occasions, "I love you." I am sure that many of you have heard God say the same. To hear the voice of God tell me in that moment that he loves me is the reassuring power that I need to survive and endure. This does not mean that I was always right in the dialogue or debate. God would convict me when my approach was unsound and when I was wrong. But even given the rebuke or correction of God in those moments, I am still able to participate in the intimate love of God, who disciplines us for our good that we may share in his holiness (Hebrews 12:10). This is intimacy—a loving and present relationship with the God of creation, who calls you his own and loves you in spite of yourself; this is the goal of life.

Knowing God in an intimate way occurs when the channels of

communication with the Father are open, available, and clear. Jesus Christ made it so through his supreme sacrifice on the cross. And now every believer has access to the throne of grace with the High Priest in Jesus who intercedes on our behalf (Hebrews 4:12–14; 12:2). Knowing and embracing this truth, coupled with the amazing reality that God journeys through heaven and earth to give us a personal word illustrates how this intimacy that is attainable to anyone who believes is priceless. This is love and the goal. Charles Spurgeon reminds us that "he who has loved you and pardoned you will never cease to love and pardon."[36] Knowing that I am not alone encourages me to live and love.

One of the most inspiring stories regarding intimacy with God is that of the Reverend Dr. Martin Luther King Jr. Dr. King recounts in his autobiography the moment he heard the encouraging and loving voice of God. In January 1956, with the Montgomery Bus Boycott just a little under a month in existence, Dr. King tells of an intimate experience with God. While his wife and newborn daughter slept, he received a disturbing telephone call. He answered the phone, and the caller, in an angry voice, said, "Listen, nigger, we've taken all we want from you; before next week, you'll be sorry you ever came to Montgomery."[37] Dr. King admitted that fear and terror consumed him in this moment to the point where he could not sleep. He now realized that he could not call on his mother or father. He needed supernatural assurance from the God. He needed an intimate touch from the Divine. Dr. King's only response was to pray. When he prayed, he heard the intimate voice of the Divine encouraging him that he was not alone. He received an answer from heaven that comforted him in a loving, sweet voice and gave him the power to fight on. In this intimate moment, Dr. King heard the voice of God

---

[36] Charles H. Spurgeon, *Morning and Evening* (New Kensington, PA: Whitaker House, 1997), 82.
[37] Martin Luther King Jr., *The Autobiography of Martin Luther King, Jr.*, ed. Clayborne Carson (New York: Warner Books, 1998), 77.

telling him to take a stand for righteousness, justice, and truth. And that voice is the one who told his disciples before he ascended into heaven, "And lo, I will be with you. Even until the end of the world."[38] After King heard the voice of Jesus, he was more charged to continue the fight for justice.

And that he did. Three nights later, his house was bombed. We all know how Dr. King and countless women and men helped change the social course of America. It is astounding, yet soothing, that an intimate moment with God provided the security to fuel his passion. There are countless examples of intimate experiences that women and men have with God. Intimate moments like these occur daily, and God desires for us to have them if we earnestly seek him, honestly expressing how we feel, because God knows and cares. God is the goal of life.

More Personal Intimacy with God

I have had moments in ministry when I have asked God if I was on the right path, track, or assignments. Working for and in the Kingdom has its fair share of challenges. Life in itself is one large continuum of challenges. But the intimacy that I share with God gives me confidence and fuels my desire to seek him. I recall one night in the winter of 2009. I was deeply distressed over the course of events that transpired in my life. I was in ministry but confused about my assignment and the direction in which God was taking me. I enjoyed ministry but was at a breaking point. I was confused, fatigued, and frustrated. I came out of a period of intense spiritual warfare in which I could not discern who was for God and who was operating against him. I had spiritual unrest. I could not sleep, and when I did sleep, I found myself wrestling again. Things were happening to me, but I could not gain understanding. I was praying to God, active in my devotional pursuit of the Holy, but not hearing his voice. The

---

38 Matthew 28:20.

silence was deafening. Thoughts raced through my mind: *Did I do something to upset God? Is God not pleased with me?* Sadly, I felt as if I was merely existing, not living. I embraced feelings of abandonment and neglect. I felt as if I was on the backside of the desert with no oasis in sight. I needed a drink from the divine well, because I was spiritually thirsty and emotionally drained. I began my routine of prayer. I was silent and felt that God was silent too. In my silence, my hope and aspiration was to tune into the airwaves of heaven, hoping to reach the correct dial. After about forty-five minutes to an hour, my world changed. I remember feeling the warm touch and a cool presence around me. I had not felt the presence of God in a while and began to ask if God forgot about me. Similar to Dr. King, I heard some extremely assuring words. My loving Father rebuked me and told me, "I am always with you; never question that!" I knew who was speaking to me, and I was familiar with that voice of comfort. I cannot even express the joy that ran through my being. The sound of God's voice and the touch of his hand revived and resurrected every dead place within me. I wanted more of him. This is just one experience of many. One of my most intriguing moments of intimacy occurred a few months prior.

On October 16, 2008, I woke up in the morning to pray. I was fasting to seek God, hoping to gain an understanding about peculiar people who had come into my life. I shall never forget that moment. As I prayed, with my prayer shoal (or *tallit*) over my head, I felt the presence of the Divine. My body began to have the same feeling of power I had when God confirmed my call in 2001. However, this encounter was more intimate. As I kneeled to pray, I felt someone walk into my room and gently sit down on my bed beside me. I understood that someone was there with me. It was like my Father came to see me. Interestingly, I did not even attempt to remove my prayer shoal or look to see who it was, because the feeling was so comforting. I do not believe that I could maneuver my head to look up or open my eyes to see. The power was so consuming, yet so consoling. I had no worry or care in the world. Everything was well. Joy and peace overflowed, and I felt that God came to visit me and was there with me. We spoke no words, but there was intimate communication—true intimacy when one can lovingly speak without words. I received a supernatural download of revelation. The revelation was the assurance that I was on my path, living out the call, and that God approved of my actions. There is no greater feeling of confidence and peace than knowing that you are in the perfect will of God and that he approves. There are deep levels of intimacy that God desires us to experience so that we can be witnesses and ambassadors of Christ and can make him known to humanity, for God is truly making his appeal through us (2 Corinthians 5:20). The goal in life is to know the one who formed us out of his innermost being.

Intimacy creates a yearning, a deep desire to be with God and to hear from him. At times the yearning is painful and you cannot find rest until you hear or sense the presence of the Divine. I have journals filled with dreams and visions, all moments of intimacy where God revealed supernatural things to me. I have yet to receive revelation for many of them. They are too wonderful for me to ever understand. Surely there are too many to even name in this book. Yet, there is one experience that I do not fully comprehend but must mention because of its nature. I had a conversation with Jesus.

My Conversation with Jesus

There are many forms of intimate prayer. Sometimes when I pray, I feel myself go into a trance. A trance is a sacred ecstasy or rapture when the mind is beyond itself, the use of the external senses is suspended, and God reveals something in a peculiar manner.[39] The New Testament word for trance in Greek is *ekstasis*, which literally means a change of place.[40] It is an act of God, initiated by God, in

---

39 S. Zodhiates, *The Complete Word Study Dictionary: New Testament* (Chattanooga, TN: AMG Publishers, c. 2000), G1611.

40 G. Kittel, G. Friedrich, and G. W. Bromiley, *Theological Dictionary of the New Testament* (Grand Rapids, MI: W. B. Eerdmans, c. 1995), 217 (translation of *Theologisches Wörterbuch zum Neuen Testament*).

which I am completely subjected to and surrendered to his power. A biblical example of the trance is recorded in Acts 10, where the Apostle Peter falls into a trance as he prays on the roof. God reveals to Peter his plan for salvation to include all men and women, from every nation and creed. In a trance, it feels as if you are literally transitioning into another reality or world.

On June 25, 2010, at about 4:40 a.m., I was in prayer, and I felt my body being shifted into the supernatural. It was the morning I was to give my first eulogy for a member at my church, for a man who was a valiant soldier for the Lord. It was also the birthday of my best friend, which we were going to celebrate that night. As I was praying, I fell into a trance, and as I went into the trance, I spoke with Jesus. I know that it was Jesus. Jesus said, "I called you to be like you are. You are peculiar." Jesus actually disciplined me. He told me, "I am not pleased with your outlook for the assignment that I called you for. You must believe and trust me." He rebuked me for a few of my prior actions.

I responded, "I am sorry, Lord, and I ask for your forgiveness." As we continued the conversation, I told Jesus, "Thank you for my parents and the love they shared. And thank you for speaking to me." Then I began to hear Jesus weep.

The Lord said, "I am sad about how my people treat me. My church is not listening to me or obeying my word." Then I transitioned back into the natural and came to my original state, on my knees. I was amazed.

I do understand how crazy this sounds, but I assure you it is true. I have yet to receive full revelation on our conversation, and there were other complex symbolisms and things in the vision that I did not mention. My point is not to scare, deter, or amaze anyone. It is to merely express the levels of intimacy that God desires for believers to have in their relationship with him. Can you imagine having a conversation with God? In actuality, we as believers do this every time we pray. We must cultivate the art of listening in our conversation with God. The goal is to walk in this level of intimacy, walking with

God as Adam and Eve did in the garden, before the fall, or like Enoch, the man who walked with God and did not die but continued to walk into eternity (Genesis 5:24).

As mentioned earlier, there are many more experiences I could name. God continues to speak to me through dreams and visions. In the majority of the dreams and visions, he or some character will reference a scripture on which I am to meditate to receive further instruction and revelation. God always points to his word. This is the mechanism of revelation that you and I have to increase our levels of intimacy. Allow the Holy Spirit to lead you into all truths through the word of God. Even as I wrote the manuscript for this book, God told me in a vision to meditate on Proverbs 10. There is a word of knowledge in the scripture that I must live by and immediately apply. But in all, the goal of the journey is the intimacy of God revealed to us in Jesus Christ through the Holy Spirit. Jesus is the seat of our affection and reason for the chase. Jim W. Goll, in his book *The Seer*, expresses the point I attempt to convey in the following manner: "After all, isn't He the goal of our passionate pursuit? He is the stream of living water that makes my heart glad! Let us pursue the "hidden streams" of the prophetic and grow in greater intimacy with lover of our soul. He is the goal of our journey."[41]

More of Jesus is the goal. God is waiting for us to discover all of these intimate moments. Again, I am only speaking from my experience, but I have discovered that many of these intimate moments are directly related to fulfilling of one's purpose.

Intimacy in the Pursuit of Purpose

Everyone has a purpose from God that he or she must fulfill. The spiritual disciplines of the faith (prayer, fasting, meditation, worship) help us to further obtain our individual purpose. Fasting from food for an amount of time or from some activity that flirts with idolatry

---

41 Jim K. Goll, *The Seer* (Shippensburg, PA: Destiny Image, 2004), 174.

helps bring you closer to God. Fulfilling our purpose helps foster intimacy, because when we began to do so, we ascertain the necessity and dependence on revelations from God to be successful in our pursuit. Some wise men once told me that you do not truly know God in an intimate way until you begin to serve him. This is the example we see in the Bible. We witness countless men and women who had intimate encounters with God because of their purpose. In fact, the Old and New Testaments are replete with examples of those who experience intimacy with God as they fulfill their purpose. From Adam to Moses, Joshua to Samuel, David to Solomon, Isaiah to Jeremiah, Mary to Elizabeth, Jesus to Peter and Paul, we all witness how intimate relationship with God bred revelation. I have personally found this to be true as well.

Intimacy with God did not begin for me until I began to seek him with more fervor, for the sole purpose of fulfilling my purpose. It was then, as I have outlined in this book, that I began to see that intimacy with God is the goal. What God does for me in intimate moments is prepare me for what is to come. There have been numerous occasions when God has given me a word of warning about how to conduct myself or concerning what will happen in corporate worship. Sometimes the night before an event, God will show me what will happen so that I will not be surprised or caught off guard. When Satan, the deceiver, plants activities to distract the people of God in Sunday worship, God has shown me what will happen and what I am to do to fight. In truth, there are times when I am unaware of what God is saying or showing me in dreams or visions. However, as the event unfolds, the light bulb comes on and the wiles of the enemy are exposed.

Whatever your purpose is, intimacy with God—the goal—will aid you in fulfilling it. Remember, God does not show favoritism (Acts 10:34–35; Roman 2:11; Ephesians 6:9; James 2:1) and is willing to pour out his love and grace on all who will accept it. I am still a sinner saved by grace, just as all believers are (Romans 3:23; 5:8). You must desire God and God alone to experience true intimacy. All who desire to see God must seek him. And it is the intimate relationship with God that will aid you in the process of learning to love others, even if you render them to be unlovable because of hurt or anguish that they may have caused.

A Way to Activate Intimacy: Love and Forgiveness

As brothers and sisters in Christ, our ultimate desire is to live our lives in such a way that we exemplify our intimacy with Christ—a mysterious oneness with God as well as with our neighbor. In doing, so we obey the greatest commandment of our Lord and Savior and fulfill the law of God: "Love the Lord your God with all your heart and with all your soul and with all your mind. This is the first and greatest commandment. And the second is like it. Love your neighbor as yourself. All the Law and the Prophets hang on these two commandments."[42]

This is the model and mantra with which Jesus, the Incarnate God, lived on earth and continues to live in us through faith (Colossian 1:27). It is easy to see how Jesus is God the Son and we clearly are not. The challenge of intimacy is to experience it with other believers when different experiences, values, and personalities are present. Yet God expects us to love one another as we claim to love God. In an intimate relationship with God, the Holy Spirit will remind you of how we must love as we are loved. This is extremely difficult to embrace, especially when you are called to love those who have hurt you in the past. The same God who loves you loves those who hurt you. They are his as well. In fact, Jesus commands us to love our enemies and pray for those who persecute us (Matthew 5:44). In this light, we become like the one we desire to love and behold. The inability to love is often married to the inability to forgive. The failure to forgive may be the culprit that hinders our intimacy with God. Not forgiving those who hurt us may be the plug in our ears that prevent

---

42 Matthew 22:37–40.

us from hearing God clearly. If your line of communication with God is distorted, check you account of forgiveness. I speak from personal experience.

Most people can tell how I am feeling by the expression on my face. I have a difficult time hiding my emotions. Yes, there are people who have hurt me in the past, and I certainly have committed my fair share of offenses. There were times when the mention of names of certain individuals would evoke irritating feelings and change my entire mood. I realized that, as I began to pray with those emotions festering in my soul, my connection or line of communication with God was distorted. Why? I reason it was because I would not bring myself to forgiveness and, thus, love. This gravely affected my intimacy with God.

The goal in life becomes more difficult to achieve because of a lack of forgiveness. I made mention earlier, in chapter 4, how a person whom I cherished at one time in my life hurt me deeply, probably unbeknown to her. At one time in our relationship, she expressed that she was not convinced that God called me to preached, at least based on my lifestyle at the time. The pain of those words drove me to seek God for affirmation and confirmation, but I held those words in my heart and resented her from that moment. The words cut me to the core, because I could not understand how a mere mortal could determine the will of God for my life.

Besides, why did it matter so much? What I was doing was for God. It took some time, years even, for me to truly face the fact that I had not forgiven her. Even after I heard from God, I carried this germ of not forgiving her. Pride was still an issue. But I had to forgive to be obedient to the command of Jesus, "For if you forgive men [and women] when they sin against you, your heavenly Father will also forgive you. But if you do not forgive men [and women] their sins, your Father will not forgive your sins."[43]

Moreover, I had to do so because it affected my relationship with

---

43   Matthew 6:14–15.

God. I had to forgive her to experience the kind of intimacy that I desired. I was reminded on multiple occasions by the Holy Spirit that I must do so. My disobedience affected my intimacy, which in turned delayed the goal. I was not expressing the love of God toward my sister in Christ. Furthermore, as long as I did not forgive or ask for forgiveness, I allowed her words to hold me captive, keeping me in bondage even though we no longer had a relationship. I did not love her the way Christ expects me to love. So I had to ask myself the relevant question the Apostle John posed: "We love because he first loved us. If anyone says, 'I love God,' yet hates his brother, he is a liar. For anyone who does not love his brother whom he has seen, cannot love God, whom he has not seen. And he has given us this command: Whoever loves God must also love his brother."[44]

None of us can say that we truly love God if we do not attempt to love one another, especially those who are not considered our favorite people. It is easy to love those who you respect and love you back. But love must be unconditional if we are to emulate Christ, the hope of glory who lives in us. By this notion alone, no one is perfect, and we will forever be in the process of becoming like the one we desire.

I had to come face-to-face with my reality. Love must be universal to truly experience the intimacy of God. This is just one of many examples throughout my life. True love and forgiveness are vital elements in maintaining real intimacy with God. It is apparent that this goal differs from the initial quest I was on. I was obviously off track. I have discovered that, as we aspire to reach the goal of intimacy that is God, life has a way of presenting distractions. Many distractions stem from prior decisions that cause us to have ties to things and people we should release. These ties hinder our pursuit. However, God is a gracious and merciful God, who will remain faithful to us even when we are faithless (2 Timothy 2:13). I have found that a certain lifestyle, at least for me, helps maintain intimacy with God, thus, further progressing the goal. It is what God requires

---

44 1 John 4:19–21.

of everyone who calls on the name of Jesus, and it helps me maintain the goal—holiness. We will deal with the concept of holiness next. Before we do, the next truth test is all about your intimacy with God. Take a moment to answer the questions thoroughly.

Truth Test

Intimacy with God is living with the notion that God loves you. It is keeping your mind focused on pleasing God with all of your actions. It is what Calvin Miller calls Christifying. It is "consciously viewing the people and circumstance in our lives through the eyes of Christ."[45] It is carrying God in your conscience with you everywhere you go. Doing this will foster communion and constant conversation with God. Take an honest reflection of your intimate relationship with him by answering these questions.

- What is your definition of intimacy? How do you view intimacy with God?
- How often do you think about God each day and for how long? Are prayer and seeking God a part of your decision-making process? Are you patient enough to listen for God to speak? Do you take time to silence the other voices when you pray by being still?
- Another way to experience God is to walk in obedience. What has God told you to do that you have yet to do? What is your apprehension or hesitance in moving on what God has said? Has your disobedience affected your intimacy?
- What or who haven't you been able to forgive? Have you forgiven yourself for what happened or what did not happen? Is not being able to forgive affecting your intimacy with God?

---

45 Calvin Miller, *Hunger for the Holy* (West Monroe, LA: Howard, 2003), 118.

CHAPTER 7

Maintaining the Goal: Holiness

Sanctify them by your truth your word is Truth.
—John 17:17

But just as he who called you is holy, so be holy in all you do for it is written: "Be holy, because I am holy."
—I Peter 1:15–16

All of my life, God has taught me what it meant to be a man of God, to truly seek him with all my heart and desire as I live for him. Receiving God's discipline and correction with humility was the catalyst that pushed me toward the attribute of God that cannot be ignored—holiness. God was making me holy. As I represent God as an ambassador of Jesus Christ (2 Corinthians 5:20), I have learned to live my life as if I love, serve, and know a Holy God. This is not to suggest that I could not achieve success in other ventures, mainly a professional career. I believe God desires that all of his children prosper. However, I have to maintain the proper focus, perspective, and attitude to what is most important in life. God has to be the priority and will not play second, third, or fourth to any person, career, thing, or idol. God is the goal!

One factor in maintaining an intimate relationship with Christ is holiness. It is what God expects and requires. There is no debate as to the character of God. He is holy. Our holy God expects his people to be holy. It is the consistent character of the God we serve. This expectation is consistent through the Old and New Testaments. As God established the law for the nation of Israel, he declared to them repeatedly the imperative to "be holy because I am holy."[46] Holy is the sense of being pure and clean in your approach to life. It is not a mere set of rules and regulations but a mind-set and lifestyle to live as God requires.

Jesus echoed the word of the Father, reiterating the importance of holiness as he gave the Sermon on the Mount (Matthew 5:48). Peter also reminded the church about the call and declaration for holiness in the scripture quoted in this chapter's epigraph. Holiness is how the angels and heavenly host address God the Father. The one attribute and character trait that the angels and heavenly host ascribe to God, giving him glory, is that of holiness.

When the prophet Isaiah recounts his call to ministry, as recorded is Isaiah 6, the seraphs (angels) who circled the holy throne of the Lord called to one another in a voice that shook the foundation of the temple, proclaiming the holiness of God: "Holy, holy, holy is the Lord Almighty; the whole earth is full of his glory."[47]

As John, the beloved disciple and apostle, awaited his execution on the island of Patmos, Jesus gave him the revelation of what was to come. When the Spirit lifted him to the heavenly realms to see the Holy One of God, John saw the throne of God in heaven. As recorded in Revelation 4, he witnessed the heavenly creatures around the throne who cried day and night in eternity: "Holy, holy, holy, is the Lord God Almighty, who was, and is, and is to come."[48]

In the heavenly realms, holiness is ascribed to God. So what does that mean for us? Since we pray in the Lord's Prayer—"Your will be done on earth as it is in heaven"—we should see that, in heaven,

---

46 Leviticus 11:44, 45; 19:2; 20:7.
47 Isaiah 6:3.
48 Revelation 4:8.

holiness in the mantra of God. If God operates in holiness in heaven, then the people of God must operate by his will, in holiness, on earth. Does that mean I cannot have wine? Does it me that I cannot go out and enjoy fellowship with friends? Do I have to be in church every night and withdraw from the world? Can I only associate with Christians and no one else? These questions may seem facetious, but they are real questions that we ask ourselves. Holiness has to do with mind-set and motivation. It is not a simple act of doing this and not doing that. It is recognizing who God is and trying to live a life pleasing to him, because you know he is watching and expects more of you. Holiness is walking and talking with integrity using good moral judgment. It has to do with why you live the way you live and ultimately understanding that you desire to please God. So what does holiness mean, and how can a lifestyle of holiness maintain the goal of intimacy? Truthfully, it is a challenge trying to live a holy life in this unholy world.

Every human is born into a sin-filled world. Sin is ever-present, and there seems to be a pull toward sinfulness. We work tirelessly to live with the mind of Christ and do the will of our God. Moreover, while noble, it is not popular to live for God in all circumstances. The truth is you need a strength and power that goes beyond human ability. To say it candidly, *you need help to live holy.*

The desire to live holy will drive you to a place of prayer and devotion. It is difficult to hold your peace when someone tries every ounce of your patience. It is difficult to remain in the perfect will of God when it seems as if those who have no regard for God are prospering in all of their efforts. It is a challenge to remain faithful to who God calls you to be when you see those who have no regard for him seemingly experience success. Since everyone else is having sexual relations before marriage and committing adultery while in marriage, why can't I? (Christians, pastors, and preacher fall into this category.) I often ask myself: *Why do you desire to please God with the way you live your life when it seems like the people you serve could care less? Who believes the Bible anyway?*

The turmoil in my spirit drives me to a place of prayer and intimacy with God where the Most High God reassures and affirms me that I am on the right path. Two years ago, I had a difficult moment in my devotional life. I felt a deep sense of uncertainty and was feeling unappreciated for my work in ministry. I was ridiculed by colleagues, preachers, and pastors for living a celibate life. I was asking God for strength, because discouragement was trying to rule my life. I was questioning my life and the critical decisions that I had made thus far. I asked God, "What is it all worth?" Then my moment of affirmation and reassurance came in a vision. As I was praying, the hand of the Holy Spirit began to write in the air. The word the Spirit wrote was *Issachar*. Issachar was a son of Jacob, and the name sounds like the Hebrew word for reward. God assured me that I had a reward for the life that I live. I cannot tell you the peace, joy, and comfort that come from having God answer your prayer and let you know that your life and the choices you make are not in vain, because there is a reward. It is the pursuit of holiness that intensifies the pursuit of intimacy.

Jesus clearly expresses the requirement and meaning of holiness in his prayer for the disciples (and therefore me and you) in the other Lord's Prayer in scripture recorded in John 17. I believe that Jesus' words to the disciples in John 17 are the key to understanding what our Lord and Savior desires of us. I constantly find myself meditating on this prayer, in particular the words in John 17:17. For me, the Christian goal of intimacy is entangled with striving for holiness.

John 17:17: The Lord's Request

John 17 is truly the Lord's Prayer. It is a defining moment in the life of Jesus because it is his last dialogue with the twelve disciples before his arrest and the commencement of his passion. To us, it shows that Jesus Christ is Emmanuel, God with us, so much so that God made the human situation his own to be with us. In John 17:17, Jesus prays to the Father, asking that his Father sanctify the disciples by the truth. The Greek word used for sanctify is *hagiazo*. It is derived from the Greek adjective *hagios*, which means holy. The literal translation for *holy* is to be "set apart." It means to be special, different, distinctive, not like everyone else, and definitely not normal. This holiness, or being set apart, is how the God of Israel, our heavenly father, identifies himself. God was not like the idols of false gods of the pagan nations. He is the true and living God who is set apart from all others. There are no imitations or replicas of God, because God is mysterious and set apart for glory. For this reason, holiness, God had to establish a law for the people of Israel. He had to reorient their minds, lifestyles, and attitudes to represent him on earth. As it is recorded in Leviticus 20:7, the God of Israel desired that his nation and children be holy: "Consecrate yourselves and be holy because I am the Lord your God. Keep my decrees and follow them. I am the Lord, who makes you holy."[49]

Although, this message is some 1250 years prior, God's chosen ones are to be holy, consecrated, set apart, and dedicated to him for his service. Jesus reiterates this to the twelve. God the Son is one with Yahweh, God the Father, and he offers the same prayer for those who are to represent him. His prayer was for God to make them holy! Since Jesus desires that we are one with him as he was one with the Father during his earthly ministry, this same prayer that he prayed for his disciples applies to all who desire to follow him. If Jesus' prayer is for me, and you are one with him and the Father, and if God is the same yesterday, today, and forevermore (Hebrews 13:8; Malachi 3:6), then what does God require of us? Holiness!

My good friend Pastor Keith Bailey reminds me that holiness is seeing as God sees. It means to have no addictions, to not be mastered by anything but God. It means to live a life with high morality (1 Corinthians 6:12–20). It means to be set apart like the God we serve. We must always remember that, at all times, we represent God. We have to have integrity in our thoughts, lives, and speech. We must

---

49 Leviticus 20:7

speak the truth and do our best to extend a hand of love. Does this sound difficult yet?

It does not mean we must be sinless, for we are all fallen creatures. Yet we are redeemed, so we are to sin less, especially since Christ lives in us (Colossians 1:27). Essentially, we should not look for opportunities to sin (premeditated). Besides, is life not filled with plenty of opportunities to sin? Every Christian must remember that he or she is a new creature with old tendencies. What helps me is a question that I repeatedly ask myself: *Self, if Jesus were looking at you, would you be ashamed?* Jesus is always looking at us and knows who we are. What are we to do?

If you attempt to live holy, you will always find yourself in a position of praying, crying for God to help you. Constant prayer will help maintain intimacy, because now you realize that you cannot achieve holiness by your own powers. You have to rely on the fruits of the spirit to help you (Galatians 5:22; Colossians 3:12). Continual prayer can only breed intimacy with the lover of our souls. What is more, the process by which we are made holy is by truth, the word of God.

The Word

In John 17:17, Jesus said, "Sanctify them, make them holy, by your Truth, your word is truth. The Word of God, the revelation that the words yields make us holy. God's Word is Truth." The words from the mouth of God as recorded in Isaiah 45:19 says, "I, the Lord, speak the truth, I declare what is right."[50] James declares that God the Father "chose to give us birth through the word of truth, that we might be a kind of firstfruits of all he created."[51] It is the word of God that is truth. Living in this postmodern society seems to be a hindrance to how people receive the word of God. Many people, even Christians,

---

50   Isaiah 45:19.
51   James 1:18.

view the Bible as just a collection of writings that hold moral and entertaining stories but have no real value of truth. This cannot be the stance of the believer who chases after God. The word of God is truth, but it is not the mere words on the page. It is the word of truth that must be revealed by the Spirit of truth. Not the words alone but the working power of the Holy Spirit that leads us into all truths.

Truth must be revealed. Sanctification and revelation are inseparable. The Trinity is one—Father, Son, and Spirit. The word comes to life through the revelatory working power of the Holy Spirit. They are not just words on a page, but when the Holy Spirit stands to work, scripture comes to life, yielding truth. Jesus encourages the Jews who believed in him to do the following: "If you hold to my teaching, you are my disciples. Then you will know the truth and the truth will set you free."[52]

I am a witness to the revelation of truth. I have a number of instances where God's word has come alive. Living under obedience and relying on God to move in the face of persecution and accusation accelerates truth's revelation. I recall one particular moment in my transition as a minister prior to entering seminary. God illuminated to me the truth of his word. Moreover, I discovered that living holy is an acknowledgment of God that, in turn, profits those who earnestly attempt to do so.

A Personal Illustration of the Holiness and Truth

My last job before I entered seminary was a challenge. I was in a supervisory position, but I was much younger than the people in my department. Furthermore, I supervised one gentleman who wanted my position, and the head of my department, with whom I worked closely, was a proclaimed atheist. I had just recently become a minister, so I was definitely trying to work out my soul salvation with fear and trembling (Philippians 2:12). To be frank, I had enemies at this job.

---

52 John 8:31–32.

No matter what I tried to do, I was constantly watched. The head of my department had access to my files, so she would frequently look to see what I was working on. She questioned me consistently. My immediate supervisor told me that she did not trust me or anyone. She thought everyone was out to get her. To add injury to insult, she had a personal vendetta against one gentleman I oversaw. He wasn't too happy about me, because I had the job he wanted. Every time I gave him an assignment or corrected mistakes in his work, he gave me attitude. Working there was a whirlwind. I was harassed on both ends and was praying for the summer to come so I could enter seminary and leave that world.

My colleagues found out that I was a minister, so then, every action was scrutinized even more. They questioned the kind of music I listened to and watched to see how I would react in pressured situations. Going to work was a hot mess. I remember on more than one occasion when another employee, one who I managed, purposefully tried to embarrass me in meetings and in public because I was new and did not have tenure. It was awful. I tried my best to remain cordial with all of them in spite of the fact that I knew they did not think too favorably of me. Because of my circus work environment, I needed a place of solace and strength. Each day at noon, I went to the bathroom stall to pray. I needed the help of the Holy Ghost to carry me through. It helped bring me to intimacy. As time progressed, I cannot say for certain that things became better, but I tried my best to maintain my disposition and motivation of holiness. I wondered if my life or presence in the office made a difference. I would discover this on my last day.

My immediate supervisor informed the office that I would be leaving and attending seminary that summer. She rallied support of the office and gave a luncheon in my honor. It was a potluck celebration where everyone in the office provided a dish, including the employee who wanted my job. Before he knew I was leaving, he had begun to lighten up, and we had a cordial and friendly working relationship. He expressed to me that he was sad to see me go because he saw that I was a man of faith. He took pictures at the celebration. I was bowled over. What is more, the head of my department gave parting words on behalf of the organization. Her last words to me, in front of the entire office staff, was that she respected my work and asked that I *pray for her* and the office as I am in school. *Wait, she's an atheist!* I thought. She did not believe there is a God, but she asked that I *pray for her*. As I stood at the head of the conference room table, which was adorned with barbeque pork chops, rotisserie chicken, honey-fried chicken wings, lasagna, shrimp lo mien, collard greens, sweet potato pie, lemon pound cake, chocolate cake, and more, her words resonated in my spirit. My Christian coworkers looked at each other with amazement. They had witnessed to her before, but to no avail. Then the Holy Spirit whispered to me the words of David: "You prepare a table before me in the presence of my enemies"[53]

This blew me away. I saw the word come alive, and God held true to his promise. I am sure you can testify to many experiences in your life when God whispered a word to you that made his word come alive. I will never look at this scripture the same, because I have an experience that signifies the truth of God's word.

I have countless examples of the revelation of truth in the word of God from my life. I could write a book simply on the testimonies of God in my life. (I share the truth of one example in the epilogue.) Through it all, holiness is the key. I invite you to try God on for size. Holiness will drive you toward him and make sure that you maintain the goal. In everything that you do, remember that God is the goal! It is God that you should desire. Believe that God desires you!

---

[53] Psalm 23:5a.

Conclusion

Intimacy with God is the goal of life. It does make sense. If we plan to spend eternity with God, then learning intimacy here helps prepare us for eternal life, when we transition from the earthly to the heavenly. What is so amazing is that, through intimacy with God, we are able to experience a taste of heaven on earth. Experiencing God intimately is the most rewarding part of life. The more you experience God, the more you desire God.

I hope that you are able to relate to or at the least are curious to know God as you are known by him. I am by no means suggesting that I have figured God out or conquered intimacy. I can only describe what I have experienced and how the Holy Spirit shapes my understanding through the study of scripture. In honesty, there are many things that I still do not understand or can't explain. I only know that intimacy with God has made the difference in my life. It is a goal that I pursue daily, because there is always more of God to pursue.

I have given up everything to follow God. As tough as it is, it is still the most rewarding journey of which I know. I cannot live without God, and I am humbled that he would consider me to be a servant. I pray that you too will pursue God and realize the goal that is God. It is the chase that provides an everlasting, eternal reward of life. Get on the chase and pursue intimacy with God. You will find purpose, and, what's more, you will discover the true, unconditional love of God. I pray that God blesses you abundantly more as you continue to chase intimacy. It is the goal of life!

No matter what we do as Christians, our goal—indeed, the goal of all things—should be to grow in intimacy with our loving Father by the power of the Holy Spirit through Jesus Christ our Lord. That is the goal of my life. —Jim Goll, *The Seer*

EPILOGUE

My Testimony of the Power of God

> "No weapon formed against you shall prosper …"
> —Isaiah 54:17

Friday, August 17, 2007, was a night that I will never forget. It was the night that my security and assurance in God was cemented, and I knew that God is real. I had a feeling that something would happen, but I did not imagine what did. Anytime I purposefully begin what I consider to be a "new assignment" from God, I have discovered that the enemy shows up. I guess the idea is to strike fear, doubt, and discomfort in me so that I will not obey God or fulfill my purpose. It is nothing more that a distraction from a defeated foe who doesn't seem to relinquish his pursuit of frustrating me. He comes to steal, kill, and destroy,[54] and on this particular occasion, the devil tried to do so literally.

I was close to two months into my relocation to Atlanta at this time. I was reconnecting with old friends from college and meeting new people as well. I was also becoming acclimated to church life in Atlanta. That evening, an old friend of mine invited me to come see a comedic, dramatic presentation at the church where his brother served as pastor. The play was good, and I hung around afterward to

---

54 John 10:10.

catch up. I recognized that it was getting late, so I made my way to my car. I was driving a rental car because I had been in an accident two weeks prior and my car was totaled. I looked at my fuel gauge, and it was almost on empty. I stopped at the nearest gas station to fill up; otherwise, I would not make it back to College Park.

It was a summer night in Decatur. The gas station was usually popular, but there were not a lot of people around this night. I thought it was strange but did not pay much attention. When I pulled up to the pump, I followed my usual routine: pop open the gas tank door, insert my credit card, and pump. But for some reason, my card was not working. I had to go inside to pay for gas, which I prefer not to do. When I came back outside to my car, I saw that the pump was finally ready and I began to pump gas. At this time, a man approached me. I was curious but not afraid. For some reason people approach me frequently to ask for directions or spare change. Besides, I normally talk to strangers and at the least say hello because I might be entertaining an angel.[55] It is just something I do, not only for biblical reasons but also because I desire to show respect for other human beings.

The man was in his early thirties and heavyset—about five foot eleven and 270 pounds to be exact. His name was Mike, and he asked me if I could give him a lift to his car. Of course I said no. He gave me a compelling story. He said his car stalled at the entrance to the highway and he needed a lift to pick up a battery or jumper cables. I asked him if he had considered calling the police to help him out, and of course he said no. (His exact words were, "Not DeKalb County Police.") I began to question, if not interrogate, him about his predicament, and he gave me some sharp answers. He told me that he was a construction worker from Columbia, South Carolina, in Atlanta on a project (which was true). We continued the conversation about his work, and it turned out that we knew some of the same people. Interestingly, he knew some people I went to high school with

---

55   Hebrews 13:2.

in Hartsville, South Carolina. I thought that this was odd, so I began to pray as I continued to speak with him.

As I spoke to him, I heard the Holy Spirit whisper, "Take him." I had never heard the Holy Spirit say anything like that before, but I was fasting to hear from God. The voice continued telling me to take him. I was unsure if this was a test from God or the enemy. It was after 9:30 p.m. Trying to be obedient to voice of God, I talked for about five more minutes before I finally decided to give him a lift to his car. He got in the car, and we pulled away. When I pulled up to his car, everything was just as he said. He was either telling the truth or a great storyteller who could lie very well. He got to the car with the keys in hand. He popped the trunk to get tools, and then we headed to toward his home, or so I thought.

As I drove, I continued praying, and everything seemed okay. We traveled on I-285, and then he told me to get off at Moreland Avenue. In the course of the conversation, I told him that I was a pastor and that my church was not too far from the gas station. He told me that his sister was a minister as well and attended a church in Duluth. He told me to make a right turn in a residential neighborhood I do not remember. We pulled up to his house, and I chose to back into the driveway, just in case I had to make a run for it. I kept the car running while Mike got out getting the battery. I popped the trunk for him to place the battery in. He got back into the car but left the door open, and had a loaded gun pointed at my abdomen.

I looked at him with disgust and said these exact words: "Come on, man. What are you doing? If you wanted money, I would have just given it to you."

He said to me in his joker-like voice, "Come up off of everything that you have. You know what this is." Then he just stared at me with a strange look.

I was scared, and my heart was pounding. But I asked him again, "Man, what do you want?"

He said, "I want it all!" I pulled out the forty dollars I had in my pocket, and then he looked at me, shaking his head, and said, "I don't want that." I was confused. A man was pointing a gun at me, indecisive about what he wanted.

"Do you want the car?" I asked. "Because it's a rental and you can have it?"

"No, I don't want the car," he said.

I then replied, "I have a cell phone. Do you want that?"

"No, what am I going to do with that?" he responded.

I finally ask him in an unpleasant voice, "What do you want?"

He turned away and then back toward me with a sinister look and said, "I want your life!"

His voice had changed. It was not Mike speaking to me, but the devil himself or one of his demon friends. I had been set up. (You probably figured that out by now.) When I heard his words, a rage from inside me came forth. I looked at Mike and made this declaration: "In Jesus' name, you want my life?" He said nothing. So I repeated myself again: "In Jesus' name, you want my life?" He did not say a word. I then began to shout with every ounce of righteous indignation inside of me, "In Jesus' name!" I must have called on the name of Jesus at least five times before I realized that I was preaching a firestorm sermon at the devil. I do not remember what I said exactly, but whatever it was, it was working. I do remember saying these words before Mike spoke: "In Jesus' name, I am a man of God, and I will not die tonight!"

As I preached, I was not looking at Mike, but he finally spoke, saying, "I need help. ... These drugs got me!" I kept preaching, and Mike began to cry profusely. I continued to preach and encourage Mike while simultaneously looking for an opportunity to grab the gun. At one point, he placed the gun in his mouth and began pulling the trigger. I began to preach to the demon and told it to leave. Mike then confessed that he was caught up, that he knew Jesus but was afraid that Jesus would not take him back. I assured him that Jesus could forgive him and that he could have a new life. As Mike cried and confessed, a man walked out to the car and told us that we needed to leave. The moment Mike looked at the man whose property we were on; I grabbed the gun and placed it by me so he could not get it. I pulled out of the driveway to get back to Moreland Avenue. I stopped the car, got out with the gun, and placed it in the trunk. I then got back into the car and began to drive and pray.

I had one hand on the steering wheel, and the other hand on Mike's forehead. As we pulled onto the highway, I told Mike to pray as I prayed. I do not remember the words of the prayer, but Mike and I went through a prayer of confession and forgiveness and I told Mike that God stilled loved him and that God can restore him. As I drove him back to the place where I picked him up, I kept encouraging him not to quit or give up on himself. I remember saying, "C'mon, Mike. Keep praying and say you believe." Halfway through our journey back to the gas station, Mike said to me that his name was not Mike but Eric. He pulled out his wallet and showed me his identification. So I then encouraged Eric to continue to pray. About fifteen minutes later, we finally pulled into the restaurant next door to the gas station where we first met. I walked into the restaurant and asked the server, "Are you a Christian?"

She said, "Yes, I am!"

I replied, "Please pray, because we have a situation where we need God to intervene."

I asked Eric, "Are you hungry?"

He said, "No, but I am thirsty." So we ordered ice tea. After a few sips, Eric began to speak and pour out his heart.

He told me that our encounter was weird to him. He said that he wanted to rob and kill me, but he could not. He said that I was protected (Isaiah 54:17). I began to ask a number of questions. How did he get into his current predicament? Why was he on the street trying to rob people? He told me everything. He was a convicted criminal who served jail time in South Carolina. That was how he knew people from my hometown, because they were incarcerated with him at one time. Working construction, he felt that he was turning a new page in his life. But recently, he was fired from that job. Instead of going home to his wife in South Carolina, he tried to survive on his own. He turned to drugs and crime. He told me that I was the third person he tried to rob. He got the car from the last couple he robbed.

I tried my best to minister to him. I explained that the path he was on would lead him back to prison. We talked for about thirty minutes. As the conversation ended, I asked him to call his wife. They spoke for roughly ten minutes; then he handed me the phone. I spoke with her and told her what happened. She said that she was embarrassed and that she told him he should come home. I agreed, and we ended the conversation. I looked at Eric, gave him ten dollars, and told him that I was taking the gun to the police.

As I was driving home, I made sure that I drove the speed limit. I could not afford to be stopped, because I was certain that no one would believe what just happened to me. When I arrived home, I immediately took the gun out of the trunk and walked inside. I tried to process what had just happened. I was in a state of shock. I lay on the bed to pray and soon felt a surge of power. It was the same power surge I felt when God first spoke to me. It was the same feeling I have when I know I am in the presence of the Lord. I wondered if after everything that happened that night, God had sent angels to minister to me. I was not sure exactly what it was, but I knew that I was in the presence of God.

As the years have passed, I have relived this moment a number of times. I wondered if it was a setup from the enemy or God's mercy operating in my foolishness. I believed with all my heart that I heard the Holy Spirit whisper to me to take Eric to his car. Was God testing me to see if I truly trusted him in all things? Was I simply naive and overzealous to help a stranger? Whatever the case, there is one thing that is certain. When the gun was pointed at my abdomen, I knew that my life was not about to end. I called on the only name that has power to help in my time of need—Jesus! Jesus saved me and kept me from death. He proved to me that the enemy trembles and is rendered powerless at the name of Jesus. After this event, the words of the prophet Isaiah to the children of Israel were literally my testimony: No weapon formed against you shall prosper!